The Engineering Student Survival Guide

McGraw-Hill continues to bring you the *BEST* (Basic Engineering Series and Tools) approach to introductory engineering education:

Burghardt, *Introduction to Engineering Design and Problem Solving*
0070121885 (GOP)

Chapman, *Fortran 90/95 for Scientists and Engineers*, second edition
0072922389

Donaldson, *The Engineering Student Survival Guide*, third edition
0072868902

Eide/Jenison/Northup, *Introduction to Engineering Design and Problem Solving*, second edition 0072402210

Eisenberg, *A Beginner's Guide to Technical Communication* 0070920451

Finklestein, *Pocket Book of Technical Writing for Engineers and Scientists*
0072370807

Gottfried, *Spreadsheet Tools for Engineers using Excel* 0072480688

Palm, *Introduction to MatLab 6 for Engineers with 6.5 Update and Additional Topics in Animation, Graphics, and Simulink®* 0072970553

Pritchard, *Mathcad: A Tool for Engineering Problem Solving* 0070121893

Schinzinger/Martin, *Introduction to Engineering Ethics* 0072339594

Smith, *Teamwork and Project Management*, second edition 0072922303

Tan/D'Orazio, *C Programming for Engineering and Computer Science*
0079136788

Additional Titles of Interest:

Andersen, *Just Enough Unix*, fourth edition 0072463775

Eide/Jenison/Mashaw/Northup, *Engineering Fundamentals and Problem Solving*, fourth edition 0072430273

Holtzapple/Reece, *Foundations of Engineering*, second edition
0072480823

IMPORTANT:

HERE IS YOUR REGISTRATION CODE TO ACCESS
YOUR PREMIUM McGRAW-HILL ONLINE RESOURCES.

For key premium online resources you need THIS CODE to gain access. Once the code is entered, you will be able to use the Web resources for the length of your course.

If your course is using **WebCT** or **Blackboard**, you'll be able to use this code to access the McGraw-Hill content within your instructor's online course.

Access is provided if you have purchased a new book. If the registration code is missing from this book, the registration screen on our Website, and within your WebCT or Blackboard course, will tell you how to obtain your new code.

Registering for McGraw-Hill Online Resources

To gain access to your McGraw-Hill web resources simply follow the steps below:

1. USE YOUR WEB BROWSER TO GO TO: **http://www.mhhe.com/donaldson**

2. CLICK ON **FIRST TIME USER**.

3. ENTER THE REGISTRATION CODE* PRINTED ON THE TEAR-OFF BOOKMARK ON THE RIGHT.

4. AFTER YOU HAVE ENTERED YOUR REGISTRATION CODE, CLICK **REGISTER**.

5. FOLLOW THE INSTRUCTIONS TO SET-UP YOUR PERSONAL UserID AND PASSWORD.

6. WRITE YOUR UserID AND PASSWORD DOWN FOR FUTURE REFERENCE.
KEEP IT IN A SAFE PLACE.

TO GAIN ACCESS to the McGraw-Hill content in your instructor's **WebCT** or **Blackboard** course simply log in to the course with the UserID and Password provided by your instructor. Enter the registration code exactly as it appears in the box to the right when prompted by the system. You will only need to use the code the first time you click on McGraw-Hill content.

Thank you, and welcome to your McGraw-Hill online Resources!

REGISTRATION CODE

SH15-GE71-1LX2-3O5R-E2IM

* YOUR REGISTRATION CODE CAN BE USED ONLY ONCE TO ESTABLISH ACCESS. IT IS NOT TRANSFERABLE.
0-07-302391-4 T/A DONALDSON: THE ENGINEERING STUDENT SURVIVAL GUIDE, 3E

The Engineering Student Survival Guide

Third Edition

Krista Donaldson

Boston   Burr Ridge, IL   Dubuque, IA   Madison, WI   New York   San Francisco   St. Louis
Bangkok   Bogotá   Caracas   Kuala Lumpur   Lisbon   London   Madrid   Mexico City
Milan   Montreal   New Delhi   Santiago   Seoul   Singapore   Sydney   Taipei   Toronto

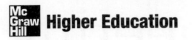

Higher Education

THE ENGINEERING STUDENT SURVIVAL GUIDE, THIRD EDITION

Published by McGraw-Hill, a business unit of The McGraw-Hill Companies, Inc., 1221 Avenue of the Americas, New York, NY 10020. Copyright © 2005, 2002, 1999 by The McGraw-Hill Companies, Inc. All rights reserved. No part of this publication may be reproduced or distributed in any form or by any means, or stored in a database or retrieval system, without the prior written consent of The McGraw-Hill Companies, Inc., including, but not limited to, in any network or other electronic storage or transmission, or broadcast for distance learning.

Some ancillaries, including electronic and print components, may not be available to customers outside the United States.

This book is printed on acid-free paper.

1 2 3 4 5 6 7 8 9 0 DOC/DOC 0 9 8 7 6 5 4

ISBN 0–07–286890–2

Publisher: *Elizabeth A. Jones*
Senior sponsoring editor: *Carlise Paulson*
Developmental editor: *Kate Scheinman*
Senior project manager: *Kay J. Brimeyer*
Lead production supervisor: *Sandy Ludovissy*
Lead media project manager: *Audrey A. Reiter*
Media technology producer: *Eric A. Weber*
Senior designer: *David W. Hash*
Cover illustration: *Daniel S. Kim*
Compositor: *Lachina Publishing Services*
Typeface: *10/12 Palatino*
Printer: *R. R. Donnelley Crawfordsville, IN*

Library of Congress Cataloging-in-Publication Data

Donaldson, Krista.
    The engineering student survival guide / Krista Donaldson.—3rd ed.
        p. cm.—(McGraw-Hill's BEST—basic engineering series and tools)
    Includes bibliographical references and index.
    ISBN 0–07–286890–2 (hard copy : alk. paper)
    1. Engineering students—United States. 2. College student orientation—United States. I. Title. II. Series.

T73.D66 2005
620'.0071'173—dc22

2004005123
CIP

www.mhhe.com

Engineering

It is a great profession. There is the fascination of watching a figment of the imagination emerge through the aid of science to a plan on paper. Then it moves to realization in stone or metal or energy. Then it brings homes and jobs to men. Then it elevates the standards of living and adds to the comforts of life. That is the engineer's high privilege.

The great liability of the engineer compared to men of other professions is that his works are out in the open where all can see them. His acts, step by step, are in hard substance. He cannot argue them into thin air or blame the judge like lawyers. He cannot, like the politicians, screen his shortcomings by blaming his opponents and hope the people will forget. The engineer simply cannot deny he did it.

On the other hand, unlike the doctor, his life is not a life among the weak. Unlike the soldier, destruction is not his purpose. Unlike the lawyer, quarrels are not his daily bread. To the engineer falls the job of clothing the bare bones of science with life, comfort, and hope. No doubt as years go by the people forget which engineer did it, even if they ever knew. Or some politician put his name on it. Or they credit it to some promoter who used other people's money . . . But the engineer looks back at the unending stream of goodness which flows from his successes with satisfaction that few professionals may know. And the verdict of his fellow professionals is all the accolade he wants.

—HERBERT HOOVER (1874–1964)
American Mining Engineer and Thirty-First U.S. President

To Mom and Dad

Brief Contents

Contents

Preface for Students

This isn't a book about how to get along with your roommate or how to balance your college budget (hey—you're an engineer, a calculator is never far away). I've tried to avoid phrases like *time management, goal setting,* and *finding yourself.* The assumption has been made that you have found yourself and an engineering program for yourself. Less lofty and (hopefully) more useful topics will be covered. Engineering students are perceived to have a heavier workload than the average student. That perception is, well . . . pretty much correct, but the perception that this occurs at the cost of other important things in our lives is simply not true.

What this book *is* about is how to learn as much as you can, get good grades, and still have fun while pursuing an engineering degree. You will find strategies to ace tests, learn to love your computer in times of cyber-crisis, land most agreeable internships, and pull through end-of-the-quarter slams in ways that are specific to engineers. All of this is not just from one engineer—but from several hundred—who have given their time and input on these subjects because they *know* where you are at.

Enough said. Much of what this book offers is common sense. So take only what works for *you* and have fun!

Preface for Professors

This is not your usual textbook. This book was written while I was an undergraduate student in mechanical engineering to provide real answers to questions about engineering and becoming an engineer to which my classmates and I could not readily find answers. While tons of guides exist for the general college student, none of them seemed to fit what we were facing in school—problem sets, laboratories, and a different academic culture than our non-techie classmates. When I wrote the first edition, support for engineering students was seemingly nonexistent or not necessarily appropriate to *students*.

As a professor, it's not easy to think back about the first day of university or the frustration of seemingly abstract math when you aren't sure of its utility. Students today are smarter, more diverse, and definitely more technically-savvy than ever before. Most of us who are now instructors have had so much engineering training that our wavelength has morphed to something on the other side of the expert-novice divide. This book tries to bridge the gap by integrating subjects important to engineering instructors and issues essential to engineering students. With each edition, this book has evolved and expanded to include suggestions from faculty and students, reviewers and readers. Literally hundreds of engineers have contributed to this work.

What's new in the third edition?

 Expanded information for minority students, particularly foreign students (come-from-away engineers) and engineering students with disabilities.

 The section describing the central engineering disciplines now includes Bioengineering.

Considerations for students taking a web-based class.

What to expect (and how to survive) a team project: forming, storming, norming, and performing.

- The addition of cartoon strips by mechanical engineering cartoonist, or as we prefer *cartoon stripper,* Jorge Cham of *Piled Higher and Deeper. Piled Higher and Deeper* humorously illustrates life as a graduate student engineer, and the younger siblings of the central characters drop in to our last chapter as they apply for grad school.

- Expanded section on writing for engineers including online resources especially for techies.

- Expert advice on *errr . . .* time management (*shhhh . . .* that was a professor request).

- Updated and pertinent information on the FE exam, GRE, salaries, engineering statistics, and a whole lot more!

Acknowledgments

I'd like to acknowledge all the folks who gave their time and effort to the creation and expansion of this book. First and foremost, *thank you* to the students who have taken the time to write to Dan or me with comments and suggestions. We receive emails from engineering students (and professors) around the world. We are always grateful to receive them—and interested to hear about different experiences. Thank you for creating a global engineering community.

For contributing their experiences and wisdom: Kathy Hannon Davies, Carla Cloutier, Ben Tarbell, John Clark, Evan Bowen, Tina Poquette, Elizabeth Loboa Polefka, Ugochi Acholonu, Brenna Hearn, Katie Otim, Tori Bailey, Sheri Sheppard, Oliver Fringer, Janay Johnson, Andy Milne, Allison Okamura, Jim Adams, Maria Yang, Michal Pasternak, Beth Pruitt—cheers! For making writing the most fun and rewarding thing in the world, even when you are writing and defending a dissertation at the same time: Kate Scheinman and her family (asante sana!), Carlise Paulson, Kay Brimeyer, and David Hash. Also thanks very much to Ken Carrizosa, Özgür Eriş, Matt Brennan, Mike McWilliams, Claire Horner-Devine, and Katherine Fringer who notably added bits and pieces. And finally an especially big thanks to Marlene and John Donaldson, Professor Bernie Roth, my 'Designer in Society' cohorts, and the talented Dan Kim and Jorge Cham.

Daniel would like to thank his parents and his wife, Woo-young, for their love and support. He hopes that someday his sons, Tayne and Taerin, will also enjoy this book. Daniel would also like to thank Krista for providing such fun and inspirational material for his illustrations.

Thank you also to McGraw-Hill's review team: Abulkhair Masoom, University of Wisconsin, Platteville; Catherine T. Aimone-Martin, New Mexico Tech; Michael M. Cirovic, California Polytechnic State University; Mary (Missy) Cummings, Massachusetts Institute of Technology; John T. Demel, The Ohio State University; Theresa M. Garcia, San Diego State University; Marguerite Hafen, Penn State University; Martin R. Kane, University of North Carolina—Charlotte; Jean C. Malzahn Kampe, Virginia Polytechnic Institute & State University; Robert G. Kelly, University of Virginia; Suphan Kovenklioglu, Stevens Institute of Technology; Ray Russell, University of Texas; Billy R. Sanders, University of California—Davis; K. S. Sree Harsha, San Jose State University; Chien Wern, Portland State University; Lisa Zidek, Milwaukee School of Engineering.

THANKS!

CHAPTER 1

Let's Take a Shot at Defining Success

Science can amuse and fascinate us all, but it is engineering that changes the world.

—Isaac Asimov (1920–1992)
Russian-American biochemist and writer

So it's all in how you define it.

Right? Sort of. As you pass through the hallowed academic hallways, others will also get a shot at defining your success with grades, friendships, and respect. Even so, ultimately *you* define your own success by setting your own expectations, limits, standards, and goals.

You've chosen your school, decided to be an engineer—or at least get an engineering education!—and even may have selected your discipline. Mom and dad are proud. Your high school or junior college math and sciences teachers are pleased. Your older (nonengineering) friends already at university are impressed with your academic commitment—"You know engineering is pretty hard. . . . Are you sure you know what you are doing?"

Engineers are admired. Engineers are cool. There is a reason:

Engineering *is* hard.

You will at times (like during the 2:00 A.M. millionth attempt at debugging a computer science assignment) curse the person who said that the college years are the best years of one's life. Many late nights aren't spent partying, but working on problem sets (while your friends may be partying). Engineering students typically have longer exams than other majors and more of them. Everything you learn builds on itself.

Of Americans who have a bachelor's (or higher) degree, 9.3 percent studied engineering.

To become an engineer (admired and cool):

You must work (very hard).

Those two points are the most important things to understand up front. The good thing is that the harder you think you have to work, the less you will realize it. Not overly comforting, eh? University isn't any less fun because you have more work to do. Think of it this way: the more work you do, the more you appreciate your fun times.

But university definitely shouldn't be all work. What else should it be?

Over 100,000 first-time, full-time students enrolled in engineering in the United States in 2000.

- A solid understanding of what was taught to you (which is hopefully reflected by your grades).

- Confidence at graduation that you are academically prepared for your next adventure whether it is in the work world or graduate school.

- Great friends.

- More "growing experiences" than you felt were needed or to which you were entitled.

- Memories to tell repeatedly to your kids—and grandkids.

- Time and opportunity to develop random interests.

The above goals may seem long range and abstract, but they come easily if you can maintain the daily LSS (Life, School, and Sanity) Balance and achieve the Ben Balance. The LSS and Ben Balances are discussed in Chapter 10, but for now that means simply that you are pleased with yourself and your surroundings on a daily basis.

So, back to success and its definition. Personal success is what makes you happy, whether it is a slick free body diagram, a perfect score on a problem set, a computer science program that *finally* works, or being able to go to bed at a decent hour before a big test. Love what you do—or at least like most of it. Engineering *is* cool.

*E*ngineering *is sometimes thought of as applied science, but engineering is far more. The essence of engineering is design and making things happen for the benefit of humanity. Engineers do basic and applied research. But all too often, engineers who conduct research are generically identified as scientists or researchers. It is no more appropriate for someone to describe an engineer as a researcher as it would be for us to depict a surgeon as a health care worker.*

American engineers endure a rigorous course of study at more than 300 accredited engineering schools in the United States for the right to be called an engineer; many continue their studies in engineering and other fields to become CEOs, university presidents, physicians, lawyers, and, yes, even astronauts. We must not dilute the value of an engineering degree by using the word engineer arbitrarily or frivolously, nor must we consign the word to relative obscurity. We, as an engineering community, must speak with pride about our engineers and our engineering achievements and not allow our profession to be wholly subsumed within the lexicon of science and technology. The distinction may be small to some, but if our children believe that engineers are relegated only to driving trains or debugging computer programs, we will have lost whatever progress we made when, 40 years ago, the Mercury 7 transformed the word engineer into hero.

(M. S.,[1] Electrical Engineering '61, Stanford University)

---

[1]The only cover to get blown: M.S. is past American Association of Engineering Societies chair Martha Sloan, a professor of electrical engineering at Michigan Technological University and the first woman to be elected chair of AAES.

CHAPTER 2

Covering Your Bases

Mary, I know what I'm gonna do tomorrow, and the next day, and next year, and the year after that. I'm shakin' the dust of this crummy little town off my shoes and I'm gonna see the world! . . . Then I'm gonna come back home and go to college and see what they know, and then I'm gonna build things. I'm gonna build airfields, I'm gonna build skyscrapers a hundred stories high, I'm gonna build bridges a mile long!

—JAMES STEWART'S CHARACTER GEORGE BAILEY
in the movie *It's a Wonderful Life* (1946)

So you are sitting in class, considering becoming an engineer. Or maybe you've known since birth that you were destined for molecules and gears. This chapter is the nuts and bolts—requirements for engineering students, getting going, and the things you need.

Let's start first with classes: To receive a bachelor's degree, engineering students must complete at least all math classes up to and including Differential Equations. To reach Differential Equations, you must complete Algebra I and II, Geometry, Trigonometry, and Calculus I, II, and III (differential calculus, integral calculus, and multivariable calculus, respectively). Some schools also require classes in Linear Algebra and Statistics. Fortunately or unfortunately (depending on how much you have paid attention in math class), the material taught in your required math classes is frequently used to explain engineering derivations, laws, proofs, and problem sets. Trig functions become second nature to the engineering student.

All engineers must have a good grounding in the sciences: physics, chemistry, and, depending on your major, biology. The better you understand these subjects *before* going into college, the better prepared you will be to handle the accelerated pace of your college professors.

> **To like engineering,** you don't necessarily have to *like* math or the sciences, but you should be confident in your ability to think analytically and solve problems.

TOOLS OF THE TRADE (OR, GREAT GIFTS FOR YOURSELF!)

What do engineers need that other college students don't? We need pretty much the same things our nonengineering peers require, except *maybe* a smarter calculator and computer, *and* both of these items can probably wait a semester or two. You will definitely need the checked items that follow during your college years and beyond. The unchecked items are useful to have but perhaps not a definite need.

☑ *Dictionary and thesaurus.* Three options now exist for wordsmithing assistance: the traditional book forms, bundled dictionary and thesaurus software (such as Merriam-Webster's Collegiate Dictionary & Thesaurus) that you install on your computer or PDA, and Internet versions **www.dictionary.com** and its sister, **www.thesaurus.com.** (The thesaurus in your word processing program hardly even deserves mention!) Each version has its pluses and minuses. Most students with access to all three find that they use all three for various reasons: the Internet may be down, you don't feel like turning on your computer, the word you're looking for isn't in the dictionary, the thesaurus has limited offerings, the bookshelf seems too far from the desk, and so on. Ultimately, you can't go wrong with a good book version. One more suggestion: if you decide to buy software, first see if your school has a site license or discounted rates.

☐ *Desk reference set.* If you just graduated from high school, it is very likely that you received one of these lovely little sets as a graduation gift. The set includes a spelling dictionary, grammar guide, random

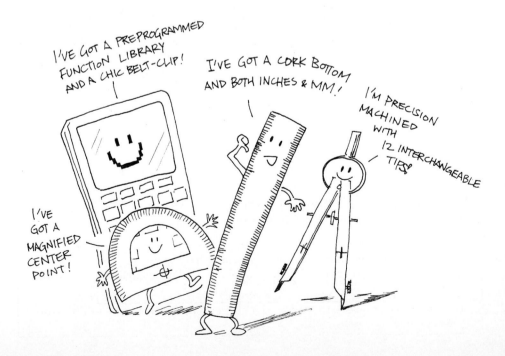

facts book (conversion tables are a definite asset), mini thesaurus, dictionary, and maybe an atlas and a few other things. Everything in the set will be useful at least once.

✔ *Paper writing/documentation guide.* Engineers still have to write a few papers! Not to worry—they are usually not long ones, and the following guides will help a lot: *MLA Style Manual* (Modern Language Association of America) or *Elements of Style* (William Strunk and E. B. White), which can also be found online at **www.bartleby.com/141.**

☐ *Compasses, rulers, other handy mathematical and graphical tools.* Believe it or not, engineers do a lot of sketching when illustrating ideas, designs, free body diagrams, best-fit curves, and flowcharts. Even a template with different sized circles and squares saves time.

☐ *Any old useful math or science textbooks or formula sheets.* In your senior year of college, you'll find you are still hauling out the freshman physics book to retrieve the formula for a sphere's surface area, or a decrepit chemistry book because you can't remember some basic stoichiometry for a combustion process calculation. Sometimes old textbooks have clearer explanations of concepts or of useful example problems. Be sure to check out Appendix D of this book, "Some Useful Stuff."

✔ *Scientific calculator.* There are two breeds of the engineer's best friend. See below.

Two Breeds of the Engineer's Best Friend
High-end calculators use two formats for computation: standard arithmetic and the mysterious RPN.

Standard arithmetic is what most calculators use. Computation is executed as it is entered and appears on the screen. For example, if you want to multiply the sum of 2 and 5 by 8, you would have to use parentheses because of the order of operations:

$$(2 + 5) \times 8$$

Standard arithmetic would allow you to enter the operands (numbers), operators (+ and ×), and parentheses in the same order you might write it.

RPN originated with a Polish mathematical logician named Jan Lukasiewicz who in 1951 wrote a book showing that mathematical expressions could be specified without worrying about parentheses. He did this by placing the operators after the operands (postfix notation). Lukasiewicz would write the above expression as:

$$8\ 5\ 2 + \times$$

While this postfix notation looks pretty confusing, many engineers swear by its efficiency. Lukasiewicz also had a prefix notation that was dubbed "Polish Notation" in honor of him. When Hewlett-Packard (HP) adapted the postfix notation to its calculators, they dubbed it "Reverse Polish Notation"—RPN—also in his honor.

☐ *Graphing and/or programmable calculator.* Wow! These things are the bomb! Graphing makes math homework sets easy to check and solve. Programming saves countless hours of time calculating roots to polynomials, solving the determinants of matrices, and so on. If you decide to purchase a really good one, wait a few semesters to talk to upperclassmen and find out what works best for you.

Whichever calculator you choose, pick one that has complex number capabilities (electrical engineers, take note!), integrates, isolates variables, in short, meets *your* needs. Suggestion: Look at calculators made by Hewlett-Packard (HP), Texas Instruments (TI), and Casio.

> ⚠️ **Watch out!** Nice calculators have been known to disappear. Carve your driver's license number on the back or personally vandalize your calculator in some other way that identifies your ownership.

Engineering is sometimes called the "stealth profession." A survey of Americans revealed that 45 percent didn't know much about engineering and another 16 percent had no clue!

☑ *Voice mail.* Engineers are notorious for their late night phone networks through which they exchange information: answer comparisons, strategies in tackling tough problem sets, and stressed and sometimes profane opinions of their major and its workload. And engineering professors are sometimes known for having trouble finding a room for help sessions, giving incorrect answers for an assigned problem set, and even postponing a test date when a student coup appears to be looming. Whether you use an answering machine or voice mail to screen calls or to save them, it gets use.

> ☞ **What About a Cell Phone?**
> Cell phones are becoming a popular choice on some campuses as an alternative to university calling plans: no dividing up phone bills among roommates and no long PINs to enter before calling Aunt Betty in Saskatchewan. There is also no yearly hookup fee or number change as you move from dorm to dorm, and a cell phone is terribly handy for tracking down missing project group numbers and directing the pizza person to the study lounge. The drawbacks, however, are worth considering, too. The expense that tends to be higher than one of the belle Bells, and the synchronous groan of your classmates the *one* time you forget to turn it off for class and someone calls are reasons enough for some to steer clear.

☐ *Computer.* Should you bring a computer? If you can, you definitely should. Owning your own computer saves you time going to and from the university computer lab, and keeps you from being at the mercy of lab hours, crowds, rules ("WHADD'YA MEAN I CAN'T BRING MY COFFEE IN HERE?!"), broken printers, and viruses. At some universities, engineering students *must* bring their own computers. However, before making any decisions about whether to bring a computer, buy one, or wait and see, check out what information your university offers on the topic. Many universities have a "Computing at *Your* University" webpage, which addresses this particular topic and covers resources available on campus (such as wireless ethernet) should you bring your own computer. There should also be information on the webpage as to the preferred types of operation systems, minimum recommended specifications (for example, ethernet capability), favored data storage (for example, ZIP drives rather than floppies), etc., in case you are the recipient of grandma's old computer. This information is important because most universities have student computer advisors to assist you—and when you need help, you want them to be able to help! If you are undecided, wait a semester or two to see whether a personal computer is a necessity. That way you will have a better idea of what model suits you best. Used computers are a less-expensive option. Suggestion: Talk to some upperclassmen!

More gifts to acquire if you bring a computer:

☐ *Backups of all software.* Bring everything, crashes happen!

☞ **Not sure what software to bring?** Not to worry—everything you will need for your classes and other activities will be available at your university. If you want to buy software or upgrade your current software, wait until you get to your school. Some software, such as e-mail programs and antivirus protection, is free to students, and other commonly used software is often offered at discounted rates. In any case, don't forget to shop around!

☐ *Software reference manuals.*

☐ *Power bar with surge protector.* Computers can protect themselves against internal power surges, but not against external surges from beyond the wall. A five- or six-dollar surge protector is cheap insurance!

☐ *Three-pronged extension cord.* If you don't use this for your computer, you can use it for the microwave or something else.

☐ *Extra phone cord* for modem hookup or *cables* for ethernet.

☐ *Antivirus, disk recovery software* for protecting your computer and your sanity. Still, check first to see if your university has a site license for any software so that students can download it for free.

☐   *Printer, paper, ink cartridges,* and a *controller box* (if you share a printer with your roommate).

 Your computer looks to you for guidance. There are three ways to be a smarter computer owner:

1. Give it a name because you will be spending a lot of time with him/her/it (you choose that, too).
2. Start a system log. This is simply a list of major changes you make to your computer—things like adding or deleting software, installing memory, reconfiguring your system, and so forth. Make note of the specific change, date, memory used or added, pertinent technical information (serial numbers, tech support toll-free numbers), and where corresponding files are located. Then, when you have a problem with your computer, it should be easy to track down and fix what's wrong.
3. If you live in a dorm, remove your computer from sight over school holidays and breaks. First floor rooms, more so than others, are targets for burglary, and computers are the most popular with thieves.

TEN TRICKS OF THE OLD-TIMERS (UPPERCLASSMEN)

You probably learned your way around during orientation week. While it is good to struggle through some things on your own, it is just as often better to ask for input. From whom? Upperclassmen.

1. Find out where the offices of the (engineering) registrar and student records are located.

If you haven't yet needed to go to either of these two places, you will in the future—to transfer credits, drop classes, order transcripts, and sort out schedule problems. When you do need to go, depending on the time of year and the proximity of the offices, it is sometimes better to call first or take a book with you.

2. Scout out the library and the computer labs. Make a note of the hours they are open.

OK, it is up to you to decide whether to use the library for studying, but you *will* be using the computer lab whether or not you bring your own computer. Many introductory engineering classes have the course laboratories in the university computer labs. Even if you can complete assignments over the network from your room, it's often easier to simply go to the lab where you have classmates to talk to (and possibly help you) and have full access to the necessary computer programs and treeware (manuals) that your class uses. It is important to note the hours of operation of the library and computer labs; if the lab closes at midnight and you have a program or assignment due at 8:00 A.M. the next morning, you probably won't want to start it at 10:00 or 11:00 P.M. the night before!

The three American universities with the largest engineering enrollments are Texas A&M (6,501), Purdue (6,096), and Penn State (5,732).

3. **Figure out where your classrooms are before classes start.**

The (often-encoded hieroglyphic) acronyms for class locations are listed in the quarter/semester schedule and, at some schools, on the Web. If your class location is not shown or is "to be announced," call the registrar or the department of the offered course. The numbering and abbreviating of classrooms and buildings can be confusing (The ground floor is the third floor? Is H530 a building or a room? Or both?). So stick a campus map in the cover of one of those stylish see-through binders and enjoy a leisurely walk around campus. When classes start and you join the crowded, 10-minute scramble between classes, you'll appreciate your foresight.

4. **Buy books early—if you can!**

Engineering books are more expensive than almost all the other textbooks for sale in the bookstore. They can put a dent in any collegiate budget and frustrate the new owner. Nevertheless, stopping in early at the bookstore ensures that you will get the textbooks you need (underordering is oh so very common), and often you will be given a choice of *new* or the much less expensive *used* books, which, depending on how early you get there, are usually in pretty good shape.

Be careful that you are purchasing the correct books for the correct section of your course. Engineering professors almost always use the same edition of the same textbook for all sections, unlike their arts and humanities counterparts, who might assign vastly different texts for different sections of the same course.

New versus Used Textbooks

University bookstores have cornered the engineering textbook market. While it may be simple enough to purchase *Renaissance Literature* or *A History of Economic Thought* at a corner bookstore, it is not so simple to find a book on biomechanical thermodynamics or transport phenomena. Used and new books don't necessarily have to be purchased from the university bookstore. They also can be purchased from upperclassmen, online discount textbook websites, and occasionally at larger schools that have discount academic bookstores.

New	Used
✔ Sturdier—important if you want to keep it	✔ Less expensive
✔ Most up-to-date edition and software available	✔ Past students have already highlighted important notes.
✔ Unblemished	✔ Sometimes answers to problems are written in the book.
✘ Higher financial loss if book isn't bought back at the end of the semester	✘ Sometimes answers written in the book are wrong!
	✘ Sometimes missing software

> ☞ **There are options to purchasing books at the university bookstore.** Nope, no illegal suggestions here. Just a heads up because engineering textbooks are expensive. Alternatives to purchasing are student-run **book exchanges** ("My thermo book for your fluids book") and **term check-outs** at the library. If you do want your own engineering library, online sellers, particularly for used books, offer lower prices. In terms of cash back for your books at the end of the quarter/semester, online sales tend to be more lucrative than the bookstore, *if* you have time.

Engineering represents 46 percent of the earned bachelor's degrees in China, about 30 percent in Sweden and Russia, about 20 percent in Japan and South Korea, and about 5 percent in the United States.

5. Find out who the deans are (and what they look like).

You could get through four or five years of education and not see or talk to the dean of the School of Engineering, but in case you should run into him or her. . . .

Deans' offices frequently offer services to students such as finding tutors, offering study courses for the licensing exam and standardized tests, and listing information on societies and engineering social events. Many universities and colleges have "pic boards" with deans, professors, and department chairs, with a list of their office locations and departmental affiliation(s).

6. Talk to upperclassmen.

Even as an upperclassman, you should utilize upperclassmen. Upperclassmen know all: which classes to take in what order, which professors to seek out and which ones to avoid, what to expect from various professors, which pizza place delivers past 1:00 A.M. to the engineering building, which electives they recommend, and so forth.

7. Be prepared for the lifestyle change.

Change is intimidating and (paradoxically) thrilling. For those who went to boarding school, the transition to college may not be a big deal, but for many others, having to share a room, actually reading a textbook, or purchasing an organizer is something new. The food won't be like home, classes are staggered throughout the day, and you may not get all the sleep or exercise you would like. And it doesn't just happen once in your four years—you'll be moving and "transitioning" maybe *twice a year* for four or five years! Finding a balance between what you *need* to do and *want* to do may not happen overnight (or even the first semester), so trust your best judgment and take good care of your health.

8. Go to departmental, library, and computer center overviews if they are offered.

Even though a lot of this information can be accessed over the Web, actually going to scheduled library and computer center orientations is key to saving time down the road locating and accessing information. Learn how library books and periodicals can be tracked down. Computer centers often have handouts on how to access the Internet, send e-mail, and figure out how much disk space students are allowed.

Departmental overviews are helpful when selecting classes and considering majors or minors. Although the thought of going to another lecture in your

spare time might not sound very thrilling, the overviews are an excellent opportunity to learn about the specific disciplines, ongoing research (could you be working for NASA your senior year?), internships, the state of the job market, and the personalities in your department.

9. Get yourself an alias.

There was a time when, if you picked *daddio, hotstuff,* or *beer4me* in a freshman fit of college revelry, the e-mail user name stuck with you your whole undergraduate career: resumes, class distribution lists, and your significant other's parents' Internet address book. At many schools, now you can have aliases—or many e-mail addresses—that bring your e-mail to one account. So now it is possible to keep your alter ego(s) and dignity!

10. Surf the TURF.

Get to know your school through an extensive web surfing session. Look for housing, maps, hours, library, student accounts, student health clinic, and everything else. Most bureaucratic stuff like declaring your major is online now. Don't waste your time standing in line at the registrar to learn that you have to go to the computer in the hall to order a student loan certificate and *then* come back tomorrow. Many university classes now have their own websites, mailing lists, and news groups. This is a great resource when figuring out classes—and workloads.

The Undergrad Engineering Experience

God help us. We're in the hands of the engineers.
—JEFF GOLDBLUM'S CHARACTER DR. IAN MALCOLM (the Mathematician)
in the movie *Jurassic Park* (1993)

The undergraduate engineering experience is unique to engineers and incomprehensible to almost everyone else. It will be one of the most rigorous stages—emotionally, physically, and academically—you will pass through in your life. Not because you are expected to learn every law, formula, and method with which you are inundated, but because you are being taught how to think analytically to become a professional problem solver (you'll hear this a lot).

Let's see, this is Chapter 3, so by now you should have a sense of what you are getting into or have already gotten into. But there is still more for you to learn! Engineering is an exclusive club. Attrition is high. Those who make it to graduation day to become engineers have talents and traits that got them there. These are inherent talents and traits, like problem solving (Before moving into your dorm room, did you mentally rearrange the furniture to maximize floor space?), resilience, determination, intelligence, and precision. You may meet a few lazy engineers, but you will never meet a dumb one.

THE TIME FRAME WE ARE LOOKING AT

How long does it take to get your bachelor's degree in engineering? *Too long* while wading through; *too short* once you've graduated. This is where the wise older professor looks over his bifocals and says: One's education *never* ends. Typically, once you have your high school diploma in hand, it takes four or five years[1] of full-time (usually nine months of the year) university schooling to complete the requirements for a bachelor's degree in an engineering discipline. The time frame may be shortened with summer school and AP credits.

Engineering course requirements, in almost all cases, are not open to any form of negotiation. So, once you decide to call the engineering buildings home, acquire two things: an advisor and a course catalog that spells out your degree requirements.

YOUR PROFESSORS

A professor's average work week is 53.6 hours: 36 hours are devoted to student-related activities (teaching, advising, etc.), 10 hours to research, and 9 hours to institutional service. (Tenured and untenured professors were lumped together.)

Engineering professors are an interesting bunch. It's hard to make any *real* generalizations about them that distinguish them from professors in other departments. Yet, once you've had a few engineering classes, it's so obvious. They are just a bit, well . . . *off*. They may laugh too loud or not at all, or they might wear polyester plaid pants with a striped shirt. More than one will be absent-minded. Some can intimidate with comments like, "My class has kept students out of

[1]Should we get more specific? According to the National Society of Professional Engineers, the length of an engineering bachelor's degree is dependent on your path:
- 4 or 5 years at an accredited university
- 2 years at a community college engineering transfer program and 2–3 years in a university engineering program
- 3 years in a science or math major and 2 years in an engineering program
- 5 or 6 years in an engineering co-op program
- 8–10 years as an engineering evening student

(The national average is close to 5 years.)

medical school." Some are complacent and use the same yellowed notes they had 30 years ago. Many are hip, caring, and remember well what it was like to be in your shoes. Others are so enthusiastic about the material that you can't help but be, too.

While engineering professors seem like a distant race of enlightened individuals, they tend to be very human—at least most of the time! If you are frustrated with class material or scheduling, the best thing to do is to go talk with them.

Professors appreciate students who come to office hours—it shows that you care! A few kindly souls have even been known to reschedule a test or due date if it falls on a day when a majority of the students have another test.

Five Things You Will Hear a Professor Say Sometime During Your Engineering Education
1. "Plug and chug" (Or "stick in the numbers and turn the crank").
2. "You kids probably don't *even* know what a slide rule looks like."
3. "And now we will show a clip of the Tacoma Narrows Bridge collapse."
4. "I took the exam myself and it only took a half hour. You guys will have three hours—it should be relatively straightforward."
5. "When I was your age, I had to run my programs with punch cards . . . and boy oh boy, if you got just one out of order and the cards weren't numbered . . . Wow!"

A Few Dos and Don'ts for Dealing with Profs

The goal is to be on the good side of every professor with whom you ever come in contact (and even those you don't). Whether you like them or not, they may be future references and colleagues.

 Don't be intimidated! You are missing out on great folks, help, and advice if you never talk to your professors.

 Do schedule appointments unless the professor has an open door policy.[2]

 Don't call professors at home. It's *not* the way to score points.

Do go to office hours prepared. Have a list of questions ready. Think how frustrating it is in class when another student wastes everyone's time. Besides dealing with us, a professor very likely has at least a handful of grad students, upcoming conferences, research projects,

---

[2]An open door policy means you can come by the professor's office any time.

engineering association responsibilities, faculty issues, and funding proposals to take up his or her time.

- **Do** always address any difficulties (the textbook could be marketed as a sleeping aid) optimistically (you are learning some important terminology, but could the professor suggest a second reference? *Note:* Just be careful your professor isn't the author).

- **Do** a quadruple check if you believe there has been a grading error on a test before you talk to the professor (see page 113). If you can, see the TA (teaching assistant) first. (A few professors have been known to dock additional points for students who complain unjustly about grading.)

ALL ABOUT ENGINEERS: CULTURE, CHARACTERISTICS, AND QUIRKS

We are a special breed (much beyond our traits and talents) that perhaps only we can fully appreciate.

The Great Divide: How to Tell Engineers from Nonengineers

What Gives Us Away	Engineers	Nonengineers
At the end of class	Watches start beeping	Rustle papers loudly and pack up their books
Calculators	Programmable, graphable, option of RPN and infrared interface	Standard checkbook style calculator—an aesthetically pleasing one
ODEs	Ordinary differential equations	Poetry, but D and E shouldn't be capitalized like that. Didn't your grammar check pick that up?
e	~2.78	i before, except after c
When picking classes	Picking classes? Hopefully, no eight o'clocks this semester	Can't be earlier than 10:00 A.M. and none on Friday
Papers	Huh?	Can pull out 10 pages in 4 hours (and still have time for spell check).
A "late night" is	Doing homework: 4:00 A.M. Going out: 1:00 A.M.	Writing a paper: 1:00 A.M. Going out: 4:00 A.M.

I know the answer! The answer lies within the heart of mankind! . . . The answer is twelve? I think I'm in the wrong building.

—PEPPERMINT PATTY
of Charles Schulz's cartoon strip, *Peanuts*

What makes engineers engineers? Programmable calculators with add-on capabilities; pocket protectors; crowded computer labs; mechanical pencils and green gridded paper; fabulous words like colloidal, thermodynamics, finite elements, eutectic, diodes, binary, gradients, flux, crystallography, eigenvectors; caffeine highs and sugar lows; goofy lab or machine-shop goggles; camaraderie and unspoken understandings.

To say that the engineering experience is about what you learn would be akin to taking a blow-dryer to an iceberg. The experience is about the people you meet, the humor you are subjected to, and the environments in which you live, work, and study. It really isn't so difficult to pinpoint a few differences that make us special (humbly so, of course).

- *Pride: Engineers know they are smart.* Of course. Rumor has it that scientific and programmable calculators are intentionally designed to look intimidating because we *like* that.

- *Stress level: Organized chaos.* There are only 24 hours in a day! So much material is covered in such a short time that it may seem disjointed and unconnected ("When will we ever use this?"). Somehow it all comes together and we manage to pull it off.

- *Humor: Quirky.* You will hear and even contribute a good share of bad jokes and puns. Hilarity will depend on your level of sleep deprivation.

- *Camaraderie: We will get through this together.* Because of our no-room-for-deviation course requirements, engineering students of the same discipline often have a majority of classes with the same group of people. This is especially true for the last 2–3 years when the classes are specific to the major or discipline. Mix this peer familiarity with the staggering workload and the result is a sense of unity from pulling together to learn from and help one another. Of course, not every school has the same workload, the same people, or even the same camaraderie, but at many places this proves true. It's hard not to bond with people with whom you have pulled many an all-nighter.

Engineers Unite (at a professional society meeting)!
An undergrad engineering experience would not be complete without participation in a student chapter of at least one of the many professional societies. Professional societies are across discipline and the gateways to life beyond campus. And do these societies have resources—wow! Just surf their webpages (see Appendix A) to learn everything you wanted to know about your future profession. Involvement with professional societies can mean regional and national conferences, local plant tours, new friends at your school and other schools, upperclassmen and professionals to tap for advice, technical competitions, scholarships, possibly a job at graduation, and much *much* more.

* *Competition: Exists.* Just how much or how little competition exists depends on your school. If you feel the chilly competitive edge of one or many of your classmates, you won't need to be told about it. There will always be one or two urchins who, on the first day, check out every book in the library remotely related to the assigned project topic. The best thing for you to do is get the books first, then share. People who are out only for themselves get the reputation they deserve.

* *Fashion sense: Generally poor.* Especially when the safety goggles go to lunch with us.

* *Social stereotype: We don't attract many members of the opposite sex.* Well, engineers may not be known for their social skills (maybe the lack of), but they are notorious for playing as hard as they work.

* *The individual: You!* Most kids grow up wanting to be doctors, astronauts, or Michael Jordan—so why did we become engineers? There are supposedly two central influences that lead students to choose engineering: The first (boringly enough) is that we were good at math and science in high school; the second, we have a parent or other close relative working in a technical field. There are other theories about qualities, likes, and dislikes that we share, such as having a Sega or Nintendo somewhere in our personal histories, a perfectionist streak, and an appreciation for "tinkering."

I don't like writing papers.
  I like problem sets.
  I like building things.
  What's this fluffy design stuff?
  —STANFORD UNIVERSITY MECHANICAL ENGINEERING SENIORS
    Paraphrased commentary on their first design class

The Female Engineer

Women have come a long way in the engineering profession, but on an absolute scale, unfortunately, that still hasn't been very far. A visual survey of the faculty pictures on the engineering prof pic board reveals that female engineering professors are few in number. They stand out as mentors and leaders, but also to greater scrutiny.

As a female engineering *student*, the unfortunate reality is you will likely face some sort of discrimination during your college career. The problem is this: *Engineering in many places is still an old boys' club.* When our folks were in school, engineering was seen as too technical, the machine shop too dirty, and the humor too crass for women. Because the professors now are mostly of our parents' generation (and older), the mind-set still exists, and women aren't always judged by their competence.

As ironic as it seems, the difficulty women face as engineering students *is* preparation for the difficult situations they will face as practicing engineers. When faced with an obnoxious comment or a disparaging remark, the best way one should handle the situation is with tact and dignity. Allow the offending speaker to see that you are above such behavior. If you are faced with a more serious situation of discrimination or harassment, contact your university women's center, a school counselor, or the ombudsperson.[3]

On the social side, a female engineer can have a lot of fun and can find being an engineer personally rewarding. Female students in engineering tend to strongly support each other and have a sense of sisterhood that other majors could never offer. Female engineers also tend to be acquainted with more of their classmates (male and female) and work better with diverse groups of people when working on projects or presentations. If you are a female engineer, the guy-girl ratio may be working in your favor, but many women opt to date outside of engineering.

Six Cool and Interesting Facts About Women Engineers[4]

1. Women engineering students tend to be more persistent than their male counterparts.
2. Female engineering students were found to be more active in their campus community than women in other majors, with involvement in honorary society leadership positions, sports teams, Greek organizations, and community service groups.
3. When thinking about transferring out of engineering, women students are more likely to seek advice from a faculty member or upperclassman than women in other disciplines. (Sounds something like the male dislike of asking for directions, eh?)

Women receive 20 percent of bachelor's degrees, 22 percent of master's degrees, and 17 percent of PhD degrees in engineering.

Nine percent of engineers in the workplace are women.

How do you fit in? The most popular engineering disciplines for women by percentage of degrees awarded are: biomedical (40%), chemical (35%), industrial (34%), materials (30%), and agricultural (30%). Least popular are electrical/computer and mechanical (both 14%).

[3]An *ombudsperson* is an impartial individual who solves disputes, investigates injustices, and strives to see that faculty, staff, and students at the university are treated fairly and equitably.
[4]See Ott and Reese entry in the bibliography for further information.

4. Most women engineering students tended to like math better than the sciences in high school, while most male engineering students tended to like the sciences better than math.
5. Research found that women undergraduate engineering students have better academic credentials than the men. In 1975 (before affirmative action), 76 percent of women engineers reported they had a high school grade point average (GPA) greater than B+ while only 36 percent of men engineers claimed a GPA greater than B+. Even more striking was that almost 3 out of 10 (28.8 percent of) women had an A or A+ average in high school compared with 7 percent of men.
6. Freshman women engineering students are younger than their male counterparts: 9 out of 10 are 18 years old or younger, while 3 out of 4 freshman male engineers are 18 or younger. Women engineering students are also younger than nonengineering women.

There are many great resources for women engineering students, but since it's not possible to mention all of them, we'll mention only a few here.

Engineering degrees make up only 2% of bachelor's degrees earned by women!

- *The Society of Women Engineers.* SWE (pronounced *swee*) is a great resource for women (and men) in engineering. Although underutilized, SWE is very active on most campuses and provides a forum for women to meet and interact with other women in engineering and industry. SWE chapters have social events such as ice-cream socials and pizza parties, and useful informational workshops for the entire engineering community on everything from internship finding to "evenings with industry" where practicing engineers from different companies and backgrounds are invited to speak. SWE's website is:

www.swe.org

- *MentorNet.* MentorNet connects women students studying engineering and related sciences (in community college, undergrad, or grad school) with male and female industry mentors over e-mail. It is free and lasts one academic year at a time, providing a framework and training for both mentors and protégés. To learn more or get involved with this excellent program, go to their website:

www.mentornet.net

- *Diversity/Careers in Engineering & Information Technology*, a free magazine aimed at women, ethnic minorities, and people with disabilities. For more information, turn to page 24.

The Minority Engineer

Twenty years ago, 98 percent of practicing engineers in the United States were white males. Current statistics show that 40 percent of engineering grads are women, ethnic minorities, and foreign nationals. While it can be tough to glean much solid information from these statistics, a glance into any engineering classroom will confirm recent trends. The problem is that we are waiting for those 20 years to catch up with the teaching staff. Women professors may be few, but African-American, Latino, and Native-American professors and instructors seem to be nonexistent. We hope that their presence in the classroom is just a matter of time.

African-American engineering students who graduated from non-HBCUs (Historically Black Colleges and Universities) give vastly different reports of their college experiences, from having excellent peer interaction and heavy recruitment upon graduation, to having less than excellent experiences with standoffish classmates and difficult group project situations. Engineering students tend to form small groups to help each other cope with the workload and difficult concepts and assignments. These groups have mostly a social basis; for example, fraternities are well known to have extensive networks of information with past problem sets on file dating back 10 years. At many schools, there is an obvious lack of cohesion and information sharing between minority and nonminority students in engineering classes (and undoubtedly others). Minority students who cope best with this sort of situation are those who get on well socially and academically with all groups.

Thirty percent of African-American engineering degrees were earned at HBCUs (Historically Black Colleges and Universities).

Six Things for the Minority Engineer to Tap Into

1. *Societies.* Minority engineering organizations like the Society of Hispanic Professional Engineers (SHPE), the American Indian Science and Engineering Society (AISES), and the National Society of Black Engineers (NSBE) are the strongest and best supported of *all* student engineering groups. These organizations and others listed in Appendix B provide an excellent network of support, mentors, and future job contacts.

2. Diversity/Careers in Engineering & Information Technology *magazine.*

   ### www.diversitycareers.com

   It's free and it's cool! This magazine is published twice yearly for all engineering minorities (women, people with disabilities, and ethnic minorities). Although the target audience is juniors and seniors with articles on topics like starting the job hunt, it offers insight for any minority engineer. If you can't find the brightly covered magazine at a society meeting or your minority engineering program office, peruse its websites. If you are looking for a job or internship, the website also offers an online resume service.

3. *Scholarships and fellowships.*

   ### www.finaid.org/otheraid/minority.phtml

   If you are worried about making ends meet, look up the National Association of Student Financial Aid Administrators (NASFAA) FinAid website or head to the library and look for directories that list scholarships, fellowships, and financial aid for minority students.

4. *History: major contributors to engineering and science.* Without much publicity, minority engineers have made remarkable contributions to research, development, invention, science, education, society, *and* engineering. Here are just a few of the groundbreakers and those who

In 2000, 5 percent of bachelor's degrees were earned by African-Americans, 6 percent by Hispanic students, 13 percent by Asians, and less than 1 percent by Native Americans.

shattered the glass ceilings changing the daily lives of everyone: Sarah Boone (African-American inventor of the ironing board), Mario Molina (Latino researcher whose work on chlorofluorocarbons led to the phasing out of their use worldwide), Benjamin Banneker (African-American mathematician, astronomer, and member of the commission that planned Washington, D.C.), Wilfred Foster Denetclaw (Native-American researcher on development of skeletal muscle), Ellen Ochoa (Latina astronaut), Garrett Augustus Morgan (African-American inventor of the gas mask and automatic street light), W. B. Purvis (African-American inventor of the fountain pen), Narciso Monturiol (Latino pioneer in underwater navigation who drove the first fully operable submarine).

5. *The future.* They've been there, done it, and made it. Minorities in industry and research are extremely supportive of you in your endeavors whether looking for a job, internship, or simply advice. Look them up.

6. *Those behind you.* Although the number of minority students in engineering programs has increased dramatically in the past decade, the overall numbers are still fairly low (see factoid to left). What can you do? Take leadership positions! Mentor other students! Get involved off campus in recruitment programs and outreach programs to get those numbers up!

Students with disabilities are role models to other engineering students—and even instructors who can't fathom engineering without all senses at maximum capacity. These sources help *dispel* the myth that disabilities and engineering don't go together:

- *EASI* (Equal Access to Software and Information) offers resources for technical professionals (and students) with disabilities. EASI's website can be found at:

www.rit.edu/~easi/

- *DO-IT* (Disabilities, Opportunities, Internetworking, and Technology) is a University of Washington program that supports students with disabilities, particularly those in sciences, technology, engineering, and math:

www.washington.edu/doit

- *The American Association for the Advancement of Science's (AAAS) Project on Science, Technology and Disability* offers several resources:

archives.aaas.org/publications
then click on "Project on Science, Technology and Disability"

- *Diversity/Careers in Engineering* magazine publishes an annual issue focusing on tech people with disabilities—their jobs, backgrounds, and lives.

The Come-from-Away Engineer

Seven percent of all engineering students in the United States are from a different country. In addition to going through many of the transitions of regular engineering students, foreign students also have to navigate a different culture, food, way of life, dress—and often language.

Resources exist in varying degrees depending on your university. Even though your adjustment to university will be tougher than most students', be assured that your classmates are also going through a life transition too. What can you do?

- *Prep yourself for the new culture.* Before you go, try to learn about social issues, customs, cultural idiosyncrasies. Seek out others in your home city who have lived in the country where you are going to university. The Web, guidebooks, and even movies will offer inside information.

- *Take advantage of your university's resources for international students and/or all orientation activities.* Experienced and friendly people can often assist you (and your family) with visa issues, settling in, and general support, like where you can get specialty food items. Orientation activities will introduce you to campus norms and slang.

- *The term* **culture shock** *exists for a reason.* Expect and accept it. A good resource on getting used to the American culture is Kansas State's Cross Cultures guide:

www.k-state.edu/counseling/culture.html

- *Aim to straddle cultures.* Stay in touch with home (e-mail is now *almost* available and affordable for everyone in the world) and seek out others of your ethnicity, but also make local friends. Both are important to your transition and settling into your new home. Volunteer activities and school clubs are great for meeting classmates (and others) in a less academic setting.

Most foreign students who come to the United States for under- graduate studies major in engineering. Interesting—the author, illustrator, and cartoonist of this book are all foreign-born!

☞ **A comprehensive guide** for students coming to study in the United States is eduPASS:

www.edupass.org

This nifty online guide covers everything from scholarships to U.S. history for foreign students!

The "Older" (Returning to School) Engineer

What makes a student "older"? Maturity level. Younger students see reentry students as the folks who ask good questions in class and really do understand things on the first pass. Reentry students are those who had a previous full-time career or job, have children/spouses, or any combination of the above. One's route to engineering can take many paths.

Students returning from other life activities to become engineers tend to have a stronger sense of mission and determination because they are making more compromises and sacrifices than the average student fresh out of high school. Professors recognize this and treat returning students with added respect, *not* because they don't respect the younger students but because, on the whole, older students are more motivated to learn and pay closer attention. Not surprisingly, they also get better grades.

Returning engineering students will not find the same flexibility in classes that other returning students may find. Balancing school and study with work and family commitments takes ingenuity and compromise. Often you can take the introductory classes in math, physics, biology, computer science, and chemistry after work at a junior or community college. After the intros, however, most necessary classes are offered only during the day, and your engineering classes will become a full-time endeavor.

If you are returning to school to become an engineer or even just thinking about it, you will have many of the same concerns and issues as any other student, plus a few more.

One in eight engineering students is over 35 years old.

- *Time for review.* If it's been awhile since you've had trig, geometry, or calculus, it is usually necessary to jump back and retake or audit[5] the last math class you took or check out a study guide to refresh your memory. The same may be true for the foundation science classes.

- *"Yikes—this is so hard."* In the beginning, many returning engineering students feel behind their younger peers and overwhelmed by the amount of work, material covered, and unfamiliar concepts they are expected to digest at what seems like lightning speed. Not to worry! By the end of your first year back, you will most likely be ahead of your classmates.

- *Most bosses are supportive of returning students.* Don't be afraid to approach them about readjusting work hours, shifts, or workloads.

- *Check out the Center for Adult Students.* Many campuses have a center or organization for reentry students that offers workshops, lunchtime seminars, and a network to meet others who are doing the same juggling act you are.

- *You will become a different person after returning to school.* You will make new friends and meet many new people; you will also have new ideas, goals, and interests. This can be alarming to someone close to you who is not in school because he or she may feel unable to grow with you. Include spouses at school social functions, take them to a favorite class (this is done all the time), and take time to show them what you are learning.

- *The married reentry student's life.* Married students who return to school may have additional issues to face, including how to support a spouse or have a spouse support them.

  1. **Work out a plan *and* a backup plan with your spouse.** Engineering is more intensive than any other program you could choose and that may be more difficult for a spouse to understand. Work-

---

[5]*Auditing* a class means you go to the class primarily to listen to the lecture. You might do the homework on your own, but you do not typically receive feedback from the instructor or official credit for the course. Schools and even professors will differ on policies for auditing a course. Some professors request that you participate in the class like a regular student, while other professors are pleased to simply have a motivated individual who enjoys their lectures. The great thing about auditing is that it often is free or inexpensive.

ing things out to a manageable level for everyone will take compromise and support from family members and/or partner.

2. **If you have children, don't be afraid to ask for and accept help.** A single or married parent who returns to school is able to do so more successfully with the assistance of grandparents, employers, church congregations, and day-care centers.

3. **More than any other student, get organized!** You are the person for whom organizers were created. Write down everything (EVERYTHING!) you can possibly remember that takes time, even dinner or going to church on Sunday.

4. **Female returning students feel a higher level of stress and guilt.** This guilt stems from their not being as available as they would like to their families or partner. If you recognize yourself here, understand that your being in school will be better for everyone in the end.

1 + 1 =__

Working parents are the fastest growing population at American universities and colleges.

Ten years ago, I was 24 years old and working full time as a legal secretary in a city 80 miles from my house. This meant a lengthy commute to work each day in traffic. Getting on the road by 6 a.m. to be at work by 8 and, then, leaving work at 7 p.m. and getting home by 9 did not leave me much free time during the day. Nonetheless, it was at this time of my life that I decided to go back to school.

I only had a high school education and I wanted more out of my professional life than working as a secretary. Not knowing exactly what I wanted to do, I decided to begin my reentry to academia with an evening calc class one night a week at the local junior college near my house. At first it was pretty crazy given my long commute and workday but I managed it. Even better, I really loved being back at school and learning again. So I spoke to my boss, telling him I wanted to continue my education and offered a deal. I would work 3 days a week for a slight decrease in pay, but I promised to still get a full 5 days' worth of work done—that way they wouldn't have to hire anyone else.

He said he'd give it a shot. So, on Mondays, Wednesdays, and Fridays, I dressed up in skirts and heels for the office, and on Tuesdays and Thursdays, I threw on a pair of jeans and spent the day at school. Yes, a little busy but, hey, variety is the spice of life, right?

I continued this routine for about 9 months when I was offered a paralegal job closer to home (160 miles of driving a day wears out even the happiest employee!). The drawback with my new job was that I needed to be at work every weekday. This definitely slowed my academic pace, but I took classes at lunch breaks. And with one here and one there during the evenings I finally reached my goal! I obtained my A.S. in Math/Science in December 1992. Yippee!

At this point, I decided it was time to take the plunge and go back to school full time—I had been guaranteed admission to UC Davis in the engineering program. I quit my job and with LOTS of financial aid, began the program at UCD. I obtained my B.S. in Mechanical Engineering in 1995.

I have not stopped since. From there, I went to Stanford where I received my MSE (Biomechanics) in 1997, and I am now a professor at North Carolina State University. If you had told me 10 years ago when I was working in that law firm that I would be doing what I am doing now, I never would have believed you. Going back to school was definitely challenging but, wow, am I ever glad I did it!

(E.G.L., Mechanical Engineering '95, University of California—Davis)

CHAPTER 4

Choosing a Major and Selecting Classes

The school should always have as its aim that the young man leave it as a harmonious personality, not as a specialist. This in my opinion is true in a certain sense even for technical schools. . . . The development of general ability for independent thinking and judgment should always be placed foremost, not the acquisition of special knowledge.

—ALBERT EINSTEIN (1879–1955)
German-American physicist

CHOOSING A MAJOR

Every major will tell you that their discipline is the best. So, undoubtedly, will the department chair at the departmental overview. Professors and "let's explore engineering" books divide the disciplines or majors into three groups: *classical* engineering (an early Egyptian had the same major), *modern* engineering (Thomas Edison invented your major), and *specialty* engineering (there aren't very many of you with the same major). However, you really need to be concerned with only two groups that are important in reaching your long-term engineering goals: *mainstream* disciplines and *narrower-focused* disciplines.

Mainstream disciplines are those that most universities with schools of engineering offer, such as Civil Engineering, Mechanical Engineering, Electrical Engineering, Industrial Engineering, Chemical Engineering, and Computer Science (which we'll consider engineering). These mainstream majors are broader in subject range and have a better selection of technical and engineering electives.

Narrower-focused engineering disciplines are tailored around and concentrate on a specific engineering field by combining knowledge from different mainstream disciplines. These narrower disciplines are often considered a branch of the mainstream majors and include fields such as Geophysical Engineering, Ocean Engineering, Forestry Engineering, Construction Engineering, and Welding Engineering. The best way to learn more about the different engi-neering disciplines is to check out school, departmental, and professional soci-ety websites (see Appendix A) or talk to professional engineers.

☞ **Curious About What Different Engineers Do?**
Check out these good sources for an overview of engineering disciplines:
- **Discover Engineering Online: www.discoverengineering.org**—This website was set up specifically as a resource to let people know what engineers are doing. Although this website is aimed at high school students, there are *many* cool things for everyone—roller coaster analysis, engineering considerations in sports and sports equipment, helicopter screensavers to download, and more.
- **Engineering Your Future: www.asee.org/precollege**—This website by the American Association of Engineering Educators has good background information on engineering disciplines as well as the practical stuff like getting money for tuition and a quiz to see if engineering is a good career path for you.
- **JETS,** formerly the Junior Engineering Technical Society: **www.jets.org**—JETS is an organization that aims to help high school students see the connection between theory learned in math and science classes and real-life phenomenon. JETS does this by sponsoring competitions and summer programs for students. Engineering Net (EngNet), which can be accessed from the JETS main page, covers topics of interest to people thinking about engineering to those just getting started. An overview of disciplines can be accessed from the EngNet page.
- **Professional societies.** All of the major professional societies (such as ASME, the American Society of Mechanical Engineers) have links to sites for students that explain what different engineers in that discipline do day-to-day on their different jobs. This may take some surfing from the home page, but it's worth it. A full list of professional society webpages is in Appendix A.
- **Skip ahead to page 37.**

Mainstream and narrower-focused disciplines are the two end points on the spectrum of possible majors, not simply headings under which every offered engineering major must be listed. It is recommended that if you are not *absolutely* certain about the best preparation for a profession in a discipline you are *absolutely* certain you will *absolutely and utterly* love, then you should pick a major that is *not* narrower-focused. However, if your potential narrow major excites you—go for it.

Here's a quick quiz that might help straighten things out if you think you might want to choose a more narrowly focused degree:

		Yes	No
1.	Did you choose your school because it offered a specific major?	☐	☐
2.	Are you returning to school after more than one year off?	☐	☐
3.	Do you consider yourself an active person with many interests?	☐	☐
4.	Did you have difficulty deciding on a major?	☐	☐

5. Have you spoken with practicing engineers
    about their profession?   ☐ ☐
6. Do you know who Mary Anderson is?   ☐ ☐

Points: 1. yes—1, no—0; 2. yes—1, no—0; 3. yes—0, no—1; 4. yes—0, no—1; 5. yes—1, no—0; 6. yes—fibber, no—the inventor of the windshield wiper and a well-known hydrogeologist.

So what does this all mean? Nothing really, *if* you are already choosing a mainstream major, but if are thinking about choosing a narrower-focused major then you should consider your quiz results. The closer your points are to the maximum of five, the more confident you should feel about choosing a narrower-focused major. If you think Construction Engineering sounds cool but you scored a two on the quiz, then take some electives in the department and look into pairing a construction minor with a more versatile degree in Civil Engineering.

Once you have started taking engineering classes, changing your major could be costly in terms of your time and money. Credits don't transfer easily the way they do for your friends in arts and sciences. So keep several things in mind when making a final decision about your major:

 A good time to specialize in a field is during graduate school. Companies that hire for a specialized area usually hire folks with graduate work in that specific field.

 Professional Engineering (P.E.) certification exams are offered only in certain disciplines. These are Civil (General, Structural), Electrical and Computer, Mechanical, Chemical, Environmental, Agricultural, Building Architecture, Control Systems, Fire Protection, Industrial, Manufacturing, Metallurgical, Mining and Mineral, Naval Architecture/Marine Engineering, Nuclear, and Petroleum. If you hope some day

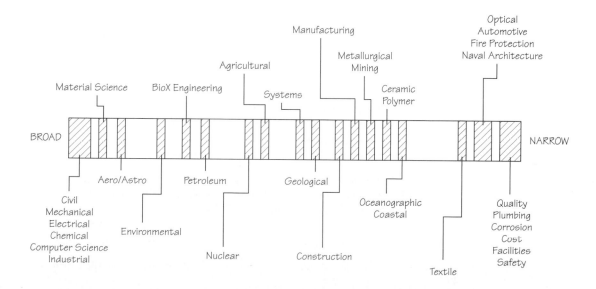

to consult in private practice or have the authority to sign off on documents and work submitted to the government, you must be a PE-certified engineer. To learn more about becoming a PE, see Chapter 11.

 The more mainstream your degree, the more versatile you are in the job market. Also, you are less at the mercy of specific technical economies and markets.

 Trust your instinct based on your life experience.

 Galileo changed his major.

THE NUTS AND BOLTS OF THE BIG SEVEN DISCIPLINES

Square One: Engineering

Before jumping into the different types of engineering, shouldn't we start with *engineering* minus the discipline prefix? That's probably the best idea considering that engineering, itself, has several definitions.

The most quoted (and most accepted) definition comes from ABET:[1]

> *Engineering is the profession in which a knowledge of the mathematical and natural sciences gained by study, experience, and practice is applied with judgement to develop ways to utilize, economically, the materials and forces of nature for the benefit of mankind.*

It reads a bit like a corporate mission statement, but that's, in fact, sort of what it is: a mission statement for your engineering training, in school and out.

Here are a few more quotes to give some breadth. Consider the similarities to the ABET definition. Notice the differences.

The words engine *and* ingenious *are derived from the Latin root* ingenerare, *meaning to create.*

> *Engineering refers to the practice of organizing the design and construction . . . and operation to any artifice[2] which transforms the physical world around us to meet some recognized need.*
>
> —Professor G. F. C. Rogers (1921–)
> *The Nature of Engineering: A Philosophy of Technology*

> *It is the Engineer who must always be the link between idea and actuality, between the probable and the practical. It is he [or she] who makes realities out of dreams. He is indeed the solvent which blends together the many different parts of our great mechanism and produces a smoothly working whole.*
> Centennial of engineering; history and proceedings of symposia, 1852–1952

> *Engineering problems are under-defined, there are many solutions, good, bad and indifferent. The art is to arrive at a good solution. This is a creative activity, involving imagination, intuition, and deliberative choice.*
>
> —Sir Ove Arup (1895–1988)
> Architectural engineer

---

[1]Accreditation Board of Engineering and Technology. For more information on ABET, turn to pages 48–49.
[2]Clever arrangement or thing.

Since there is no one standard definition of engineering, this book will take some liberty in adding another by cutting and pasting common ideas and using a bit of engineering slang:

> *Engineering involves using technical skills and scientific principles (gained in school and through practice) to create <u>elegant</u> solutions to serve the needs of the planet.*

Maybe you've heard *elegant* used before in this context, or in a similar one related to problem solving. An elegant solution is creative, yet practical, efficient (in terms of time and resources), astute, and some might even say *beautiful* in its simplicity.

Engineering Work (and Passion)

You've undoubtedly heard stereotypes about engineers. Nontechies (also called "fuzzies") may say that engineering is cold, austere, and passionless, but nothing is further from the truth for the engineers pushing the realm of scientific knowledge, applying that knowledge, and making dreams possible. Well-known engineer and inventor David Levy reported bursting into tears the first time he saw someone, who didn't also work for Apple (he asked first!), using the Macintosh Powerbook he had helped design. The stranger was using the laptop in a way that Levy and other designers had anticipated.

Not all engineers do only design work, however. If you think about any research or science project you've done already, you know that there are several stages to a project before reaching a conclusion or outcome. At large companies like Boeing or DuPont, imagine *your project* several orders of magnitude bigger—more research, more ideas, and more people working on solutions. There, the engineers may be specialized—some doing research, some doing design, some testing, and some providing customer support.

At startups, on the other hand, there may only be a few engineers—so one engineer takes on many different tasks that, at a larger company, would be divided among several engineers. An engineer may have done part of the early research, designed the implementation, and helped write the users' manual. Let's briefly consider the stages of a project and the engineering work associated with each. You've probably noticed that although different project tasks exist, they may not necessarily occur in a chronological order. The process of engineering development may be cyclic or zigzag forward and backward before the final conclusion is reached.

- **The seed.** The seed is the impetus for your project. The seed typically evolves from an idea that leads to a question or need. For example, in wondering how a radio works, you may ask the *question:* Why is the reception better when you stand near your radio but worse as soon as you walk away? A *need* (think like a buyer!) may be a domestic-use radio that has good reception regardless of its location in a room.

  A seed may originate inside or outside the engineering organization. When the seed comes from customers or potential customers (outside the company), it is called a *technology pull* because technology development is spurred and subsequently occurs to meet the customers' need(s).

On the other hand, new technologies also spawn from advancing science or even unexpected finds. In this case, companies may have access to a cool new technology, but are unsure of how to apply it or what to apply it to—so they may try different variations of the technology to address existing needs. This is called a *technology push*.

Engineers doing work at the seed stage are business and market oriented, trying to discover consumer needs or other unaddressed gaps that exist in current technologies and processes. They work with other corporate departments (like marketing and research) to evaluate competing and existing technologies and the risk of new ventures. Can a new technology fulfill a need at an affordable cost?

If all the Play-Doh made since 1956 was extruded through the Fun Factory®, it would make a "snake" that would wrap around the world almost 300 times. The weight of the snake would be more than 700 million pounds— that's equal to the weight of almost two Grand Coulee Dams (Washington)![3]

Can You Guess What Viagra and Play-Doh Have in Common?
Viagra and Play-Doh are both examples of technology pushes that were able to respond to existing market pulls. Viagra was originally developed by Pfizer Pharmaceuticals in the early 1990s as a heart medicine for men. Viagra wasn't very successful in reducing chest pain, but it did have another *interesting* side effect that was able to meet a consumer need.

Kobul Industries manufactured soap and cleaning products. One of their products was a nontoxic wallpaper-cleaning compound, but in the 1950s, the demand for wallpaper cleaner was decreasing with the improvement of home heating systems. Joe McVicker needed to do something to keep his family's Ohio plant in operation. Play-Doh's story began with a conversation between Joe and his sister-in-law Kay Zufel, a nursery school teacher, who mentioned that the modeling clay supplied to her class was too firm for little fingers. The lightbulb went on, and Joe, with some colorful dyes and assistance from other family members and co-workers, came up with Play-Doh.

Research. Once you have a seed, it's time for some research. The level of research involved in engineering projects varies greatly. On one end of the spectrum, an engineer may immerse himself for a few days in a topic with which he isn't familiar to explore design options and potential solutions. To design a better radio, an engineer may need to learn about "smart products" if he wants a digital interface. At the other end of the scale, a full-time research engineer may work in a laboratory analyzing material and processing properties to develop smaller circuitry than what exists. Her work may lead to a digital radio the size of a pin head.

Engineers who work full-time in research are at the forefront of development. They study and expand on the knowledge that currently exists in search of better and different solutions. They tend to be patient, focused (it can take a while to get useful results), meticulous, and comfortable with the abstract.

Design. What are possible answers to the posed questions? What are potential solutions to a need? How might technologies be arranged or rearranged to provide solutions? A good designer considers even the wildest ideas in trying to figure out how to turn ideas into reality. How might a useful new technology be integrated with an old technology to produce a radio that works well—no matter where it is located? Can it be done at a reasonable cost? Ideas may start as sketches and evolve into mock-ups out of cardboard. Form and function radio prototypes are made out of foam and simple circuit boards. Slowly with work, input from potential consumers and other engineers, testing, and iterations, these early prototypes become more and more sophisticated until the final technology results.

Full-time design engineers have good spatial visualization skills (they can picture and manipulate shapes in their mind[4]). They like thinking "outside the box" to come up with novel (elegant!) solutions to problems. They are also the "tinkers"—always trying to figure out how things work and how they might work better.

Analysis. Analysis involves calculations based on engineering theory and experience. Time for the gridded green pad and mechanical pencil? Probably not. The industry mediums of choice are spreadsheets and modeling software or specialized codes. An engineer working on the wristwatch digital radio may do thermodynamic[5] calculations and simulations to determine if one circuit component may cause another to overheat. Occasionally, experts are temporarily brought onto engineering teams to do specific analysis that is beyond the scope of the team's skills.

All engineers do analysis in their daily work, most simply by thinking critically: "Does this make sense?" Engineers who most enjoy analysis tasks have a good grasp on how the physical laws of nature and engineering theory relate to reality: "Cooool. It works the way it is supposed to!"

Trusting Your (Engineering) Intuition . . .
Engineering intuition is the ability to be able to tell if numbers or solutions make sense in the context of the problem. Knowing a number is too large or small isn't necessarily something you are born with—you can pick it up with a bit of practice. How? Just think about how your calculated results compare to what you know about reality. This kind of intuition is invaluable to engineers because it enables them to do "back-of-the-napkin" calculations to do quick preliminary analysis.

Implementation. Numbers have been crunched, a solution has been designed, but there is still application. Engineers working in implementation are the ones in the plant making sure production occurs

---

[4]If you weren't born with spatial visualization skills, it can come with practice!
[5]Having to do with heat and energy conversions and transfers.

*Wondering how
your tuition
contributes to your
university's
operation? On
average, it makes
up 18.8 percent
and 43 percent of
revenue at public
and private
universities,
respectively.*

as planned or on site making sure that the technical questions are addressed appropriately.

Whereas a design engineer may be in charge of turning the idea into reality, it is often the engineer involved with implementation that makes reality possible. If you decided to mass-produce your radio, a manufacturing engineer would be the right one to tell you how to get production up and running. He or she would be able to tell you which product attributes are suited for which manufacturing processes. An industrial engineer might help you organize and synchronize the whole production system.

These making-it-happen-everyday engineers spend a good deal of time outside of their "cubes" (cubicles)—talking to machinists, foremen, vendors, and other folks who can make recommendations about bettering the implementation of a technology. They are problem solvers, spending time troubleshooting, improving the overall system, or checking back with other departments to facilitate necessary redesign.

Testing. Calculations have suggested how a new technology is supposed to behave, but do we really know? Not until we test it. Testing is crucial—to the planet and the product. Technology can improve the lives of people, but it can also ruin their lives if not properly tested. Testing must also show that the engineers' work stands the test of time and has no detrimental effects on the environment.

In the testing stage(s), technologies are "maxed-out" or pushed to the limit, not always in the most obvious ways. Engineers doing testing need to anticipate all the unintended effects and usage the technology might see. Could the wristwatch radio interfere with a plane's navigation system? Sometimes they need to be creative to simulate conditions that are impossible or unsafe to test. These engineers must be very observant, precise, and detail oriented; the smallest data glitch can highlight serious design flaws.

How to spot an engineer in a crowd. In case there are no ink marks on the breast pocket or chemical burns in one's fleece pullover, you can always check out one's *hands*. Most engineers in Canada and many in the United States wear a ring on the pinky finger of their "working hand." The "engineering ring" symbolizes not only our pride in the engineering profession, but also reminds us of the importance of a high standard of ethical conduct. In the United States, *The Order of Engineer* ceremony asks recent graduates or practicing engineers (at some schools you will see some graduates initiated with a practicing engineer parent!) to accept the Obligation of the Engineer and wear a stainless steel ring. In Canada, where the tradition originated, the secret ceremony is called *The Ritual of the Calling of the Engineer* and initiates are given an iron ring to wear. Early rings were reportedly created from the remains of the Québec Bridge (made of iron), which collapsed in 1907 killing 82 construction workers. Investigators of the tragedy concluded that the fundamental contributing factor was engineering design errors.

For more information on both organizations, go to:

Order of the Engineer: **www.order-of-the-engineer.org**

The Ritual of the Calling of the Engineer: **www.ironring.ca**

Sales and Customer Support. Many high-tech sales and customer support personnel are engineers. To be able to knowledgeably discuss the benefits of and answer important questions about the workings of a technology, you must be able to understand its capabilities and limitations.

Engineers who do sales and provide customer support are good communicators. Think how difficult it might be to explain how to hook a DVD player up to a television *over the phone* to someone who isn't familiar with "this kind of thing." A job working with customers may offer a lot of traveling to trade shows and client sites.

Time for some engineering reflection! Stop reading right now and look around the room. Everything (EVERYTHING!) you see requires engineering: design, testing, development, troubleshooting, research, and support. Even a can of soda or the pencil you are using now—consider the finer attributes of each that make them user friendly, safe, functional, and a mainstay in our daily lives.

And that is only the tip of the iceberg—engineers also work in a variety of other jobs. Many become CEOs, managers, lawyers, doctors, astronauts, writers, and teachers (among other things). The possibilities are endless.

The Big Seven: The Nuts and Bolts

Engineers most commonly describe their jobs and expertise by discipline. This is similar to other professionals (otolaryngologists[6] and patent attorneys, for instance) who also start their education learning the foundations of a field then progress to specialty areas.

Although engineering is segmented into these various disciplines, a great deal of overlap exists. For example, if you wanted to take the Introduction to Robotics class, depending on professors' interests and the size of your school, there may be many robotics classes offered through several departments. It makes sense—real-life problems rarely fit neatly into one framework. Many universities are recognizing this and encourage students to do multidisciplinary undergraduate work to better prepare for jobs. So, for instance, if you are interested in robotics and medicine, the robotics class offered in the Biomedical Engineering Department would likely best suit you. Take advantage of the overlap!

To add one more complicating note to the labeling of disciplines, subdisciplines of engineering may be categorized differently depending on the school. Manufacturing at one school may be in the Mechanical Engineering Department, but in the Industrial Engineering Department of another.

The Big Seven: Civil and Environmental Engineering, Mechanical and Aeronautical Engineering, Electrical and Computer Engineering, Industrial Engineering, Chemical Engineering, BioX Engineering, and Material Science and Engineering are briefly overviewed in the following section. Look deeper into any one of these disciplines and you will see it is impossible to comprehensively characterize the work of any engineer. Many more disciplines than

[6]Ear, nose, and throat doctors.

the Big Seven exist, but they overlap with and/or draw on material from one or several of them.

Engineering for the Citizens of the Planet: Civil and Environmental Engineering

Although the term *civil engineering* was first used in the eighteenth century to refer to "civilian" work done by military engineers, civil engineering is the oldest of the engineering disciplines (or at least leaving the most long-lasting artifacts!). Think about ancient pyramids, aqueducts, and religious buildings—their structural design and construction are mind boggling, even by today's standards.

Civil engineering in the present day encompasses engineering activities having to do with construction, transportation, structures (like bridges, buildings, and dams), soil stability, waste management, and the environment. Environmental engineering grew out of sanitary engineering, a branch of civil engineering, with the integration of ecology, hydrology, microbiology, chemistry, hydraulics, and other environmental sciences with engineering during the mid-1960s.

"Civil" comes from the Latin word civis, *meaning* citizen.

Civil and environmental engineers work in the public sector for international, national, and regional environmental and aid organizations and agencies; health departments; and municipal engineering and public works departments. In the private sector, they work at consulting engineering firms, construction contractors, utility companies, and industries.

What kind of work might you find a civil or environmental engineer doing before heading off to lunch today?

- Getting weekly drinking water samples to test for EPA (Environmental Protection Agency) compliance.

- Talking with an architect and project manager via conference call regarding the second floor design of a small hospital.

- Asking an implementation question at a workshop about new landfill lining technologies and methods.

- Developing a model to predict the rate of increased vehicle traffic and expected travel patterns on major roadways in a medium-sized urban center.

- Developing a flow release schedule for a dam to meet environmental and power generation constraints.

- Monitoring the spread rate of an oil plume from a leaking tanker.

- Using software to analyze weather patterns to find more fuel-efficient flight paths for airplanes.

- Flying over a golf course and condominium construction site to investigate developers who may have not followed water management permitting regulations.

Engineering for Smooth Movers:
Mechanical and Aerospace Engineering

Mechanical engineers are called the "general practitioners" of engineering because they typically know a little bit about everything engineering—particularly if it relates to the concept of *motion*. This includes the analysis and design of energy and fluid systems, machines, consumer products, and *much* more. Aeronautical and aerospace engineering is closely related to ME, but focuses primarily on the aspects of engineering related to flight of human-made things.

Because of their broad training, mechanical engineers can find a job in almost every type of organization that requires technical and engineering skills. The largest demand for aerospace engineers comes from industries that support military and general aviation, but aero/astros are also hired by other industries equally interested in streamlining, fluids, and controls expertise.

What might you find a mechanical or aerospace engineer doing before lunch on Tuesday?

- Delving into thermal analysis of computer CPUs[7] to determine what size fan, or fans, will be needed to keep them within operating specifications for a given temperature and altitude.

- Debugging software that automatically adjusts satellite altitude for optimum signal transmission.

- Performing supersonic wind tunnel tests on fighter jet engines to minimize noise due to shock waves and turbulence.

- Designing a heat sink (to dissipate heat more quickly) for a welding jig that was producing slightly distorted parts.

- Flipping through the Springs-N-Things catalog to find an appropriate leaf spring for a new mouse prototype.

- Calculating possible spacecraft trajectories to model (and minimize) operating costs on a spreadsheet.

- Manipulating flow analysis software to minimize the drag induced by the side-view mirrors of a new line of sports cars.

The Manipulation of the Invisible:
Electrical and Computer Engineering

Electrical engineers are fundamentally interested in the electron—how it moves, its relation to other charged particles, and most simply the application of electricity. This isn't the electricity of Ben Franklin, but a field that includes physics, chemistry, manufacturing, electronics, communications, optics, and computers. Computer engineering is closely linked to electrical engineering, but focuses on all aspects of computer hardware and software design, construction, implementation, and monitoring.

Electrical and computer engineering are the largest engineering disciplines; they make up approximately 25 percent of the engineering workforce.

---

[7]Central processing units, the main processing chip of a computer.

*More engineering
students at
American schools
are registered in
electrical and
computer
engineering
(113,653) than any
other discipline.
Second
is mechanical
(65,115).*

While you were checking e-mail, an electrical or computer engineer was:

- Developing circuitry for a laser to be used for eye surgery.
- Writing image-processing software that animators will use to put animated movies together.
- BLEEEEP[8] for a secret telecommunication satellite project.
- Programming a chip that reads signals to monitor minute changes in ambient lighting.
- Analyzing a power grid in a section of your city to minimize ineffi ciencies.
- Doing some quick calculations to size needed power supplies for the testing department.
- Performing efficiency and cost calculations for a parallel computing system to make a supercomputer in a government lab.

> ☞ **What's the Difference Between a Computer Engineer and a Computer Scientist?**
>
> Both computer engineers and scientists have hardware and software skills. The CS major will typically know more about the software—coding, algorithms, programming, and computing models—whereas the computer engineering student will be more focused on hardware—computer architecture, circuitry and logic devices—and design of specific components to meet application needs.

The Earliest Chemical Engineers Weren't Necessarily Making Root Beer

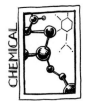

Chemical engineering reportedly dates back to the introduction of process industries, suggesting that the earliest chemical engineers were the fermentation and evaporation experts of ancient civilizations. Modern chemical engineers were distinguished from chemists in the last century with the application of engineering skills to the design and development of high-volume chemical manufacturing operations. A chemical engineer, however, does more than oversee plant operations; he is concerned with rates of chemical processing (evaporation, distillation, filtration, absorption, extraction, etc.), robust designs that minimize the effects of external constraints (like temperature), the economies of raw materials, and the processes required to manipulate material behavior.

Most chemE's work in chemical and oil industries, but they can also be found in a wide range of other process industries, such as the food and paper industries, electronics, biotech, advanced material, and health industries.

A smattering of things a select group of chemical engineers were trying to finish up before taking off for the weekend include:

- Cleaning up the lab following preliminary testing of fire-resistant synthetic fibers that will later be as comfortable to bare feet as carpet.
- Leaving a voice-mail message for another engineer that the two lubricants he had left a querying message about earlier in the day were indeed compatible.

[8]Censored.

- Making a few notes from discussions about a resin that can be used to control the release of active ingredients of medical tablets.

- Answering a student question about water-in-oil emulsion following an industry seminar: "Chemical Engineering and Personal Care Products" at the local university.

- Finishing up a written analysis of by-products produced in the initial stage of the manufacture of a new fertilizer.

- Making a list of potential application methods for a polyurethane adhesive that will be used for food packaging.

Keeping Industry Industrious: Industrial Engineering

Industrial engineering is the field of engineering that supports implementation by using scientific principles to efficiently manage systems involving people, production, equipment, and materials. The need for mass production methods (particularly around World War I) was the official beginning of industrial engineering as a field of its own. Today, almost 75 percent of industrial engineers work for manufacturing companies, but many work in other organizations varying from health care providers to government agencies.

An industrial engineer came in a bit early today to work on:

- Preparing her monthly quality report based on the statistical data on high-tolerance, injection-molded components for a cell phone.

- Developing a spreadsheet that calculates ride efficiencies (how long the lines might be for a ride and how many people will ride in an hour) at a theme park that will be opened next year.

- Configuring software to track the inventories, work-in-process, and orders of a company's popular line of backpacks.

- Analyzing forecasted data from the marketing department to decide how much to increase production and inventories for expected order increases before the holiday shopping season.

- Talking to workers in the riveting department to find ways to reduce muscle strains and noise level.

- Sketching a block diagram on a large pad of paper to conceptualize the overall functional design needed for an automation system that will lift, transport, and release long sheets of metal for stamping.

- Reconfiguring a product's manufacturing layout since 400 square feet of plant space is going to be turned into a new employee lounge.

Sports Equipment to Superconductors: Material Science and Engineering

Material science and engineering is the study of the behavior of (usually solid) materials based on their composition and structure. The closest relatives to this field are solid-state physics, metallurgy, and chemistry.[9] Material engineers

[9]Some university curriculums will allow a substitution of a material science class for a chemistry requirement (it depends on your major and school!).

manipulate material properties for a wide range of engineering applications in every area of technology development. Specialty areas exist in metals and alloys, ceramics, polymers, composites (a combination of materials), and, more recently, semiconductors.

The last one hundred years have produced an explosion of new materials and jobs. Nylon was commercially developed only 65 years ago! Like a product, materials are designed keeping in mind application needs, ease of manufacture, and environmental safety, among other things. Material engineers work in a variety of industries enabling technology to continue its rapid advance.

While you are in math class, a materials engineer may be:

- Removing a material from cryogenic conditions to observe microscopic structure rearrangements.
- Reviewing data from the latest tests of a new alloy to be used for engine blocks.
- Meeting with a biomedical engineer and doctor to discuss the design of a new medical device to assist dysfunctional kidneys.
- On vacation explaining to his 8-year-old daughter how her composite skis can bend so much without breaking.
- Writing out specifications for a low-weight, high-temperature ceramic tile to protect spacecraft during reentry.
- Running down to the lab to check on some samples of recycled thermoplastics that were dropped off yesterday.
- Meeting with his team to find ways of reducing manufacturing costs for IC (integrated circuit) wafers.

> **A fab Internet resource** for more info on all aspects of material science and engineering is the Career Resource Center for Material Science and Engineering:
>
> **www.crc4mse.org**

The Techie Side of Life: BioX Engineering

BioX engineering encompasses all the bio- <u>fill in the blank</u> disciplines of engineering, such as bioengineering, biomedical engineering, and biological engineering, and subdisciplines such as biomechanics, biomaterials, and so on. This exciting new field bridges the gap between engineering and the life sciences. BioX engineers are interdisciplinary folks, trained typically in engineering fundamentals (primarily electrical and mechanical engineering), biology, and other appropriate areas such as anatomy, environmental sciences, and medicine. The history of this field has many roots: the environmental side of bioX engineering in practice goes back to the origins of farming, and the human side can be traced to biomedical engineering, which got started in the 1950s. Biomedical and related engineering fields resulted as a response to obvious needs within medical fields for individuals with engineering expertise to develop medical instrumentation and assisting devices, model the mechanics of the body, and research materials acceptable to the body.

With the expansion of biotechnology fields, bioX engineers are in high demand to enable more efficient and effective research. They work in industry for pharmaceutical, medical device, environmental, and health systems companies. They also work in hospitals, at government agencies doing defense work, testing, or policy administration.

What might you find a bioX engineer doing at 10 A.M. on a Thursday?

- Debugging the user interface of a 3D computer modeling program that targets where radiation should be delivered to cancer tumors.

- Being interviewed by a reporter who is writing an article on the "roto-rooter" device he designed to remove plaque from clogged arteries.

- Discussing with a surgeon a biocompatible "bone cement" that can be used to increase the density of real bones at artificial hip implantation sites.

- Fiddling with a do-hicky from a prototype valve, which is part of a kit for a more efficient hydroponics[10] system.

- Reading a technical paper on preparing the optimal corn fiber oil that will lower cholesterol.

- Searching through Genbank for the DNA sequence data of a bacterial protein that might assist in "eating up" contaminants.

"Bio" comes from the Greek word bios *meaning* life.

What Will the Next 100 Years Bring?
No one doubts that engineering is the discipline of the future as humans become faster, smarter, and more efficient. How can we state that confidently? Just look at the last one hundred! The National Academy of Engineering in collaboration with engineering organizations and professional societies selected the 20 engineering achievements that they believed to have the greatest impact on the quality of life in the last century.

The 20 Greatest Engineering Achievements of the Twentieth Century:

1. Electrification	11. Highways
2. Automobile	12. Spacecraft
3. Airplane	13. Internet
4. Water supply and distribution	14. Imaging
5. Electronics	15. Household appliances
6. Radio and television	16. Health technologies
7. Agricultural mechanization	17. Petroleum and petrochemical
8. Computers	18. Laser and fiber optics
9. Telephone	19. Nuclear technologies
10. Air conditioning and refrigeration	20. High-performance materials

For more information or specifics of each, go to the website:

www.greatachievements.org

Copyright © 2000 by National Academy of Engineering.

---

[10]The science of growing plants in fluids without dirt

*Imagine trying to
pick classes and
sections at one of
these schools! The
top five American
colleges or univ-
ersities by total
enrollment levels:
1. Community
College of the Air
Force (AL) 63,700;
2. The University of
Texas—Austin
(TX) 48,900;
3. Miami-Dade
Community College
(FL) 8,500;
4. Ohio State
University—Main
campus (OH)
48,200; and
5. University of
Minnesota—Twin
Cities (MN)
45,410.*

MINORS, DOUBLE MAJORS, HONORS DEGREES, AND OTHER THINGS

An added distinction to your degree is always nice. It offers you more flexibility in your job search and offers employers a more versatile potential employee.
Check if there are special programs or minors available at your university for
engineering students. Some curricula in one discipline include so many classes
from another that a minor or honors distinction may be only an extra two or
three classes away. Discuss your options with your advisor.

Except for electives, engineers aren't so much able to select classes as they are able to select *sections*[11] of classes. Upon notifying the university that you will be attending their esteemed institution, you are likely to receive a handbook with a course schedule similar in detail to the one given on pages 46–47 (especially if you chose ME).

Hmm . . . not a whole lot of room for flipping through the course catalog and selecting random classes that sound interesting. What can you do to make a rigid schedule less constraining? Here are 10 tips to make your schedule and time work for you.

1. **Don't overburden yourself the first semester of your freshman year.**

   Just as you didn't learn to ride a bike overnight, so you will need to ease into college life and its academic demands. Allow yourself a lighter load first semester and expect some bruises. How well you do your first semester could set the tone for your remaining semesters. You will learn your limits by pushing them—and others' expectations of you by sometimes not meeting them. A strong academic first semester lays a solid foundation on which to build your remaining college years; a less-than-strong first semester can dig you a hole from which you must climb out before you start on the foundation.

2. **Check out the professor before you register for his or her section.**

   A professor who bounces around the classroom on a pogo stick to demonstrate kinetic and potential energy will not only keep you awake and entertain you, but engrave the concept and application in your mind. Your like or dislike of a professor and his or her teaching methods makes a huge difference in your interest and success in that class, your enthusiasm for the subject, and even your confidence in choosing engineering as a career. Upperclassmen are an excellent source of information when selecting classes.

Is Your Professor Live . . . or Televised?
Some Pros and Cons to Taking the Televised Class at Home

Pros	Cons
✔ Professor is often an excellent lecturer	✘ Paying attention can be tough
✔ Convenient	✘ Easy to tape and never watch
✔ Classes can be taped and watched a second time or important parts rewound and rewatched	✘ Feeling of detachment; may be intimidating to seek out TA
	✘ No interaction during lecture—questions have to wait

[11]A section is a subdivision of a course. For a class like Introduction to Chemistry, there might be 300 students in your lecture, but only 15 in your lab *section*. In Calc I or II, there may be 300 students taking the course, but you would only see 15–20 students in your section. Different sections are usually offered at different times, and sometimes have different professors.

☞   **Taking a Web-based class?** Several eco-friendly professors have moved to paperless homework and notes: all course materials and homework submissions are online. Although there seems to be mixed feelings about this model among students (ranging from "I just print everything out anyway!" to "I love having all the course material there and organized!"), there is consensus that because of computer glitches and web traffic, online homework submissions should *not* be left until last minute!

Specimen Curriculum for Mechanical Engineering

FRESHMAN YEAR

		Semester Hours	
		Fall	Spring
Chem 102a,b	General Chemistry (I & II)	4	4
Math 172a,b	Analytic Geometry & Calculus (I & II)	4	4
	Elective	3	
ES 130	Engineering Science Modeling and Simulation	3	
Physics 117a	Elementary Physics I		4
CS 101 or 102	Programming and Problem Solving		3
	Total	14	15

Note: Materials Science 151 (Introduction to Materials Science) is a 4-hour course that may be substituted for Chemistry 102b.

SOPHOMORE YEAR

		Semester Hours	
		Fall	Spring
Math 222	Analytic Geometry and Calculus (III)	3	
Physics 117b	Elementary Physics (II)	4	
ME 160	Mechanical Engineering Modeling	3	
CE 180	Statics	3	
	Elective	3	3
Math 229	Elementary Differential Equations		3
ME 190	Dynamics		3
EE 112	Electrical Engineering Science		3
ME 171	Instrumentation Laboratory		2
ME 220a	Thermodynamics I		3
	Total	16	17

JUNIOR YEAR

		Semester Hours	
		Fall	Spring
ME 200	Kinematics	3	
ME 212	Engineering Mechanics Laboratory	2	
ME 213	Energetics Laboratory		2
ME 220b	Thermodynamics II	3	
CE 182	Mechanics of Materials	3	
ME 201	Design of Machine Elements		3
ME 244	Fluid Dynamics		3
	Elective	6	9
	Total	17	17

Note: ME 212 and ME 213 may be taken in either order.

SENIOR YEAR		Semester Hours	
		Fall	Spring
ME 202	Design Synthesis	3	
ME 263	Heat Transfer	3	
ME 257	Engineering Systems Analysis		3
ME 203	Senior Design Project		3
	Electives	9	9
	Total	15	15

Elective Breakdown and Requirements: Social Sciences Electives = 9 hours and Humanities Electives = 6 hours (at least 3 hours in each must be at 200-level), Math (above 229) = 3 hours, Technical Electives = 9 hours, Technology and Society Electives = 6 hours, and Open Electives = 6 hours.

☞ **A famous or accomplished professor isn't always the best educator.** Ask around about his or her teaching style and availability to students.

---

3. **Take classes in the recommended sequence.**

This seems fairly obvious, but . . . a lot of schedules are ideally arranged so that concepts taught in one class will carry over into other classes taught at the same time. For example, from our sample curriculum schedule (see page 46 and above), you can test the concepts from CE 182 Mechanics of Materials first semester junior year in the ME 212 Engineering Mechanics Laboratory experiments. Although the course guide states that labs ME 212 and 213 do not have to be taken in a specific order, you should take classes so they complement each other. It will be much less work for you. This matters at some schools more than others.

4. **Be realistic—don't take an eight o'clock if you won't get up.**

For some people, getting up for an eight o'clock class is no problem, but lectures at two o'clock on Friday afternoons (especially when ski season arrives) are another story.

5. **Fill course requirements early.**

Get them out of the way! If you think taking *another* math class again next semester is torture, try taking it second semester senior year when you can't remember *anything*. Are depth, breadth, and technical electives being met with the courses you have selected? Double-check with your advisor. A mistake could force you to attend summer school.

6. **Don't schedule more than three classes back to back in one day.**

Have you ever heard of the Attention Span Quota? Probably not, it's made up—but it is still very real. Active listening and critical thinking can be taxing on the ol' noggin, so give yourself a break by breaking up your schedule. Otherwise, you may exceed capacity and find yourself paying more attention to the school newspaper crossword puzzle than your lecturer.

7. **Be wise in choosing nonengineering electives.**

These are the classes that fill the Renaissance-student requirements needed for your diploma. They can provide some much-needed stimulation for the left side of the brain, but recognize that engineers already carry a heavy workload.

Approximately 14 percent of college freshmen who declared engineering as their major need remedial work in math, and about 8 percent need remedial work in the sciences (biology/physics/chemistry).

Thus, you should be careful what you choose or you'll find yourself struggling with time and frustration.

> ☞ **Having trouble deciding between two electives?** See if the professors have class webpages, or next time you are at the bookstore, browse through the textbooks for the classes.

Some Considerations for Choosing Nonengineering Electives

BEWARE of classes that . . .	CHOOSE classes that . . .
✘ May be too reading-intensive (literature survey classes).	✔ Are discussions oriented.
✘ May be research intensive (like Readings in Old Norse and you don't speak or read Old Norse).	✔ Are recommended by other engineers.
	✔ Are unique to your university or are taught by famous professors.
✘ Have substantial added fieldwork, language labs, or films in addition to regular class time and homework.	✔ You've always been intrigued by.
✘ You aren't enthusiastic about.	✔ Are a change of pace from your required classes for the term.
✘ Have prerequisites.	✔ Improve your communication skills for a future job.

Does this mean you shouldn't take a random class that looks interesting because it will demand a bit more time? Definitely not, but evaluate how the workload from it will fit in with your engineering workload, extracurricular commitments, and social schedule. Although it would be fun to take an elective in scuba or advanced web design, you'd probably get more out of the course if you took it during the summer when more time is available.

> ☞ **Ahem, one last thing.** You also might want to avoid taking any classes that you don't want future references, employers, or graduate schools to see on your transcript (e.g., Modern Sexuality and Eroticism). Of course, you can always opt to sit in on the course.

> **Who Decides What Engineering Students Learn?**
> It is your university that ultimately decides which material, in what order, is appropriate for someone in your major. But then—you are probably wondering—how can graduates from several hundred engineering schools all qualify for the same job? And who makes sure that your professors are teaching the material you need and are supposed to learn? The answer is ABET (Accreditation Board for Engineering and Technology), a national organization that officially accredits university engineering programs. The criteria ABET uses for accreditation is based on the input of the 31 professional engineering and technical societies that comprise ABET. Universities are not required by law to be accredited, but some states require that you graduate from an ABET accredited program to be eligible for the FE/EIT (the first step in becoming a licensed engineer—see Chapter 11). Universities do not receive institutional accreditation; ABET reviewers (many practicing engineers) evaluate and appropriately accredit each discipline separately.

In the 1990s, evaluation entailed reviews of the qualifications of the faculty, the adequacy of teaching facilities (including laboratories, equipment, computer resources, and libraries), and the work done by the students. Each university program needed to meet or exceed the minimum requirements established by the profession to receive accreditation. However, a new and improved set of criteria (Criteria 2000) was recently introduced, which moved away from the traditional "bean-counting" sort of measurements. The new criteria state that graduates of accredited programs must be able to meet measurable standards not only in technical proficiency but also in communication skills and knowledge of engineering in a greater social and global context. With the new criteria, universities must state how their students will achieve these things, and how they will measure it.

For more information (unfortunately, it's pretty technical) or to check to see if your school is accredited, go to ABET's website: **www.abet.org.**

8. Look into taking nonengineering pass/fails.

It's the best of both worlds: getting credit for a class without having to do a lot of work. Pass/fail (sometimes also called credit/no-credit or credit/no clue) policies vary from university to university and even from department to department, so check with your advisor to see if pass/fail is an option.

9. Register as early as possible for classes.

Putting off registering or forgetting to do so until later can strand you in a Monday evening chem lab that directly conflicts with prime-time football. Registering early also gives you a better chance of getting a better professor—open classes and sections with good professors offered in late morning and early afternoon slots fill up quickly. Also note cutoff dates to drop, withdraw, and add a course—failure to do so can result in annoying paperwork or a late fee.

10. After freshman year, lighten your second semester or spring quarter loads.

A lighter load does not necessarily mean taking fewer hours, but perhaps lightening up other commitments; for example, cutting down on extracurricular involvement or part-time job hours (if financially possible). Christmas vacation and spring break are usually not enough to recharge most students' internal batteries. The way to fight fatigue is to plan for it. See tips on fighting burnout in Chapter 9.

Approximately 1/8 of all the classes taken by an engineering student is math.

> *I think the best course I took was an experimental course on entrepreneurship. I saw it announced on a bulletin board that I used to scan for interesting events. It was listed as a graduate level course (I was a freshman), but I went to see the Dean and I talked him into letting me take the course and report back to him about how appropriate it might be for an undergraduate. The instructor had started several businesses of his own, and he gave a first-rate course. He only taught it that one year, then went on to other projects. The course opened up a whole new world for me, inspiring me to get involved in a startup venture when I graduated.*
>
> *(A.M., Electrical Engineering '91, Penn State)*

CHAPTER 5

In Class—More to Staying Awake Than Taking Notes?

It is through science that we prove, but through intuition that we discover.
—Jules-Henri Poincaré (1854–1912)
French mathematician

It can take until junior year before you really feel like an engineer in your selected discipline. Your freshman and sophomore years are spent getting the background, the skills, and the tools that you will need for your major: math to understand the proofs and derivations, computer skills to set up both analytical and illustrative models, and confidence in your own best methods of analysis and problem solving. Engineering, as you will read in this book and hear many times at school, builds on itself. All that math you take during the first years of college is *frequently* used later on, not only in the sophomore, junior, and senior years, but also in summer internships, graduate school, and your life as an engineer. Notes from a freshman computer science programming lab may be hauled out and dusted off to find the command to solve your senior year design project.

Unlike many of our nontech peers, we do need to absorb as much as we can from *every* class to ensure a strong basis for the next level of learning. During a particularly hard week, sitting still in a marginally comfortable desk for more than 10 minutes might fool your brain into thinking it's time for sleep. Your manipulation of the classroom environment and your preparedness for the course offer the most buoyancy in these tidal-wave times. In this section, strategic ideas are offered on note-taking, paying attention, getting more out of classes and laboratory periods, and recognizing the important elements of a course that will assist you in conserving time and energy and establishing some good(!) habits early on.

In Class—More to
Staying Awake
Than Taking Notes?

*11.8 percent of
college students
think that the
chairs in
classrooms are
comfortable.*

One of your classrooms might be a small stuffy room with a noisy heater, another might be a huge lecture hall with hard, tiny seats and constricting desktops. The psych major will inform you of the salutary wall and ceiling colors of the psychology building and McDonald's decor, but unfortunately *that* train of thought was derailed before pulling into the engineering buildings. OK, so there may not be a whole lot you can do about your classroom to encourage paying attention to the professor, but you *can* control where you sit. Keep these things in mind before grabbing the seat closest to the door.

 Sit smart. Believe it or not, studies have shown that the closer a student sits to the front of the class, the better that student does in the course. The details are sketchy, and it seems a tad odd that someone would research this subject, but take it for what it's worth. Consensus seems to be that you should sit near the door if you are late (so you won't disrupt the class), in the back if you want to nap or do other homework—otherwise wherever you are comfortable.

 Is your professor a leftie? For classes in which you copy a lot of notes from the chalkboard, sit on the left side of the classroom if your professor is left-handed and on the right side of the class if your professor is right-handed. This helps with professors who seem to write faster than 3×10^8 m/s (speed of light), but perpetually eclipse the chalkboard.

 Call it seat security. The seat an engineering student sits in the first day of the semester or quarter is often the one to which he or she returns each time that class meets. If it isn't the same seat, it is often in the same area of the classroom. Comfort through familiarity?

NOTE-TAKING AND PAYING ATTENTION

Not only are they related, but *note-taking and paying attention are entwined.* Taking good notes may get you good grades even if you never look at them. How? Because to take good notes, you must listen actively to your professor. Active listening means digesting the information and making the connection between the theories and the applicable examples. Of course, you should frequently look over notes from your classes, but even then, isn't it much nicer to go over complete, neat, informative notes that are properly organized in chronological order? So for EXCELLENT notes that lead to MOST EXCELLENT grades:

1. **Listen actively!**
    For the reasons explained above. Here are a few hints for maintaining alertness in times of trouble:

First Period (the 7:00 or 8:00 A.M. class)	• Eat a healthy breakfast (a cereal called Sugar Bombs may sugar bomb you in the middle of class). • Get up and do something nonsedentary before class—go for a quick jog, run an errand. • Could you secretly be a morning person? Try going to bed and getting up to study *before* class.
The Drowsy After-Lunch Period (1:00 or 2:00 P.M. classes)	• Eat lunch after class or somehow readjust your meals so that postmeal drowsiness doesn't set in while you are in class. • Eat lunch somewhere you are forced to walk or bike to class (the exercise may help). • Sleep in, go to class alert.
Tuesday and Thursday Periods (Or any other unusually long class)	• Save and use your nice pens so note-taking is kind of fun. • Don't sit anywhere near friends or someone you will want to distract or talk with. • Fidget! It's good for you (see page 141) and it will keep you awake! • Be prepared for class. You'll find it becomes much more interesting.
Last-Class-of-the-Day Period	• Almost there! Just get through this period. • Bribe yourself. • Remind yourself that you want to learn this only once.

2. Use binders and keep a hole-punch in your backpack.

Halfway through the semester the wire coils on spiral notebooks are bent, the pages won't turn, and those notes you photocopied a couple of weeks ago have vanished. Some people just aren't binder people, but engineers should be—it's so logical. Class handouts can be immediately integrated with notes (thus the hole-punch) and binders allow for neat note rearrangement for studying convenience.

Binders can be divided to include more than one subject (the Monday-Wednesday-Friday binder) or all your subjects to cut down on the weight in your backpack and the probability that a book is forgotten at home.

3. Date everything, including handouts.

Most people do it anyway. It makes things so much easier at exam time and when discussing class or a problem with your classmates over the phone ("But look in your notes and see what he said on January 26 about thermoplastics").

4. Copy everything the professor writes down or reiterates.

Engineering classes are not usually discussion oriented. So, if a professor writes something on the chalkboard, it is probably important for you to comprehend or at least refer back to later for help with problem sets. If the notes on the chalkboard are nearly verbatim to the course textbook, still copy them down. Copying and listening will keep you focused *now,* saving time reviewing the text *later.* What about the professor who forgets about the chalkboard until he illustrates a picture from a "really good" engineering anecdote? Make note of what topics he hits and study the textbook.

5. Leave gaps on the page, not in your notes.

Pages that are broken up with bullets, headings, and diagrams are much more fun to read (easier anyway) than pages of compressed writing. Is there a paper shortage? Leaving spaces on the notepaper also allows for later study additions or relevant notes (see number 6).

6. Use two colors of ink.

A nifty idea is to use one color for what the professor writes on the chalkboard and a second color for your own supplements to the professor's notes. The supplements can be clarifying comments or examples the professor mentions, definitions in the text, or even questions to ask during office hours. The benefit of using the two colors is to be able to distinguish what the professor deems important from your supplementary information.

☞ **Use inks that won't run or bleed.** If getting caught in a downpour isn't bad enough, losing good notes will worsen your mood considerably. Waterproof ink (ballpoints and some felt tips) also allows for highlighting without bleeding.

---

7. Box, underline, or highlight important equations, useful definitions, and theories.

All the equations begin to run together when there's a proof involved or even when "simplifying" another equation. If your professor deems anything important: box it, star it, highlight it. Somehow make it jump out at you from the rest of your notes because it *will* appear later on a test. Guaranteed.

8. Employ a user-defined (your own) shorthand.

It saves time. Some professors will write so much that they recycle their own written-out notes from year to year as overheads just to fit it all into the class period. Your brand of shorthand also will keep your hand from cramping and save your brain from tedium. Use apostrophes! Use abbreviations! As long as you can understand it—use it.

Math Stuff

Δ	change		#	number
(+)	positive		∞	infinity
(−)	negative		//	is parallel to
=	is equal to		$\perp$	is perpendicular to
$\neq$	is not equal to		i.e.	that is
$\equiv$	is defined by		e.g.	for example
$\cong$	is approximately equal to		cw or $\curvearrowright$	clockwise
$\sim$	approximately		ccw or $\curvearrowleft$	counterclockwise
$\overset{?}{=}$	might be equal to		✌	peace
$\rightarrow$	leads to		DEF	definition
$\uparrow$	increases		$\measuredangle$	angle
$\downarrow$	decreases		<	section
$\Sigma(..)$	sum of (...)		$\Pi(..)$	product of

Taking-Notes Stuff

FS	factor of safety		abs	absolute
impt	important		cont.	continued
rel	relative		b/c or ∵	because
w/	with		w/o	without
∴	therefore		eqn	equation
mat'l	material		sol'n	solution
			hmk	homework

Q.E.D. Quod Erat Demonstrandum—which was to be demonstrated, used at the end of math proofs.

9. Title examples

Yes, it sounds rather odd. However, if a professor takes the time to go over an example, usually this problem illustrates something important. Chances are this "something important" will appear later in a problem set, a test, or an exam. Determine what makes each problem unique from the others discussed in the lecture. Classifying each example with a summarizing and distinguishing name helps the mind neatly file the method away and easily recognize a similar problem at test time. How would you title the example fluids problem?

An Excerpt from Some Lecture Somewhere . . .

Your Fluids professor says:

"OK—to illustrate flow over a flat plate normal to the flow, let's consider the force on a billboard exerted by an 80 mph wind during a hurricane. Remember that wall shear stress won't contribute to the drag force in this situation. The billboard is 30 feet by 10 feet with a thickness of 1 foot. The wind is blowing normal to it. We'll assume the kinematic viscosity and density of air to be 1.8×10^{-4} ft^2/s and 0.00237 slug/ft^3, respectively." (She is drawing on the billboard as she talks.)

continued

You write:
(Space for title)
V_{wind} = 80 mph, blows ⊥ to sign
Dim'ns of sign: w = 30 ft, h = 10 ft

ASSUMPTIONS: $v = 1.8 \times 10\text{-}4 \frac{ft^2}{s}$, $\rho = 0.00237 \frac{slug}{ft^3}$

You draw:

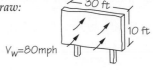

V_w=80mph

Fluids Professor:
    "Converting the wind velocity to feet per second, we have 117 ft/s. Now calculating the Reynold's number with the sign *width* we get 1.95×10^7." (Also writing the equation on the board.)

You add:

$$Re = \frac{Vw}{v} = \frac{(117 \text{ ft/s})(30 \text{ ft})}{(1.8 \times 10\text{-}4 \text{ ft}^2/s)} = 1.95 \times 10^7 \qquad \underline{\text{USE WIDTH}}$$

Fluids Professor:
    "So what does our Reynold's number tell us? We'll use Figure 9.10 on page 442 because our Reynold's number is greater than one thousand. To use the chart we need a b-to-h ratio, so for us that is billboard width to billboard thickness which is equal to 30. Reading off the chart we maybe have a drag coefficient of perhaps 1.6."

You add:
$Re \geq 10^3$ <span></span> <u>use FIGURE 9.10 on page 442</u>

$$\frac{b}{h} = \frac{w}{t} = \frac{30 \text{ ft}}{1 \text{ ft}} = 30 \quad \rightarrow \quad \text{Chart gives } C_D \approx 1.6$$

Fluids Professor:
    "And finally now we can use our Drag Equation. Plugging and chugging, our correct answer is 7786 pounds."

Your notes:
(Space for title)
V_{wind} = 80 mph, blows ⊥ to sign = 117 ft/s
Dim'ns of sign: w = 30 ft, h = 10 ft

ASSUMPTIONS: $v = 1.8 \times 10\text{-}4 \frac{ft^2}{s}$, $\rho = 0.00237 \frac{slug}{ft^3}$

$$Re = \frac{Vw}{v} = \frac{(117 \text{ ft/s})(30 \text{ ft})}{(1.8 \times 10\text{-}4 \text{ ft}^2/s)} = 1.95 \times 10^7 \qquad \underline{\text{USE WIDTH}}$$

$Re \geq 10^3$ <span></span> <u>use FIGURE 9.10 on page 442</u>

$$\frac{b}{h} = \frac{w}{t} = \frac{30 \text{ ft}}{1 \text{ ft}} = 30 \quad \rightarrow \quad \text{Chart gives } C_D \approx 1.6$$

$$D = \frac{1}{2} C_D \rho V^2 A \qquad \underline{\text{Drag eqn.}}$$

$$D = \frac{1}{2}(1.6)\left(0.00237 \frac{slug}{ft^3}\right)\left(117 \frac{ft}{s}\right)^2 [(30 \text{ ft})(10 \text{ ft})]$$

$$= \underline{7786 \text{ lbf}}$$

After copying this down, how would you characterize this problem? What's important in this example? What equations do you use? What distinguishes this problem from other problems?

Some good titles might be:

- Wind drag on sign
- Sign problem
- Drag on flat plate normal to flow
- Gone with the wind

> ☞ **A Note on NOTES.** Save everything. No cooking s'mores over engineering notes, you'll *need* and *use* them again.

HOW TO SUCK THE MARROW OUT OF CLASS

This section might more stiffly be called "How to Get Your Tuition's Worth" or "How to Maximize Your In-Class Learning and Understanding." Here are a few things to consider that will help you cut some corners on the map of time (and exam-time cramming).

1. Bring what you need to class.

> ✹ *Yourself!* Let's apply some of our analytical engineering thinking to demonstrate the importance of good class attendance.

Three Choices: Choose Your Own Adventure!

Go to Class—Tired, but alert	Go to Class—Zzzz (You had a Biology test first period)	Skip Class (You go back to bed after the bio test first period)
Skim over recommended reading. *= 20 min.*	Go to class and snooze through about half of it. Notes are so-so. *= 50 min.*	Zzzz at home in bed. *= wasted tuition*
Grab a soda or coffee after the first period test. Go to class, pay attention, take good notes, and understand class examples. *= 50 min.*	Read all of recommended text to fill in holes in notes—What was important? *= 1 hr., 45 min.*	Track someone down to get notes. *= 10 min.*
Look over notes later and read excerpts of recommended reading that are important. *= 40 min.*	Unsure why width is used instead of height in class example. *= no time, just annoying*	Go get notes. *= 15 min.*
		Photocopy notes and wait in line to pay. *= 20 min.*
		Return notes. *= 15 min.*

The Harris survey "American Perspectives on Engineers and Engineering" found that many parents would be very pleased if their children became engineers. When asked the question "Using a scale of 1 to 10 with 1 being extremely displeased to 10 being extremely pleased, if your son or daughter or other family member said they wanted to be an engineer, how pleased would you be?" The mean response was 9.

continued

Go to Class—Tired, but alert	Go to Class—Zzzz (You had a Biology test first period)	Skip Class (You go back to bed after the bio test first period)
	Call classmate to find out why width is used instead of height. (Hey, does anyone have a student directory I can borrow? Wait, what's her last name?) *= 15 min.*	Read over notes twice, but have difficulty deciphering handwriting. *= 55 min. +* *frustration*
		Read all of recommended text. I *think* this section is supposed to be important. *= 1 hr., 30 min.*
		In passing, learn from a friend that a different section is important. *= added frustration*
		Read again and comprehend the important section. Work the text example problems. *= 1 hr.*
		Phone friend to ask why width is used instead of height in class example. Chat. *= 15 min.*
= 1 hr., 50 min.	*= 2 hrs., 50 min.*	*= 4 hrs., 30 min.* *+ tuition wasted* *+ frustration*

And these are the *best* scenarios. Notice from this example that it would still be better to go to class even if you might not pay attention (you will have a brainache some days) than it would be to skip class completely.

What else should you bring? That depends on the class and the professor.

March 14th is pi day.

- *Textbook.* Some professors will frequently refer to tables, charts, and examples in the textbook. So bring your textbook and note important topics and problems.

- *Straightedge, compass, graph paper, etc.* Bring any tools that will assist or speed up your neat note-taking; a university bookstore mini stencil speeds up drawing flowcharts or molecular configurations. A ruler (or student ID) makes for nicer graphs and diagrams.

- *Ministapler.* It's handy to have with you when problem sets are due every day or every other day. Having a stapler also makes you popular with your classmates.

Hole-punch. Leave the industrial-size punch at home; a handheld one is sufficient (see page 53).

2. Avoid chatty classmates and entrancing posters.

That's interesting . . . [strain to read the poster] 3-D COMPUTER-AIDED MAP-PING OF THE ADIRONDACKS. Nifty! . . . I wonder if my wilderness brochure came today. Hmm, gosh, I wonder if I got *any* mail today? Uh-oh, forgot to mail my credit card payment again.

Avoid anything that distracts or interferes with focusing your six senses on the professor's lecture. A conversation behind you can make it difficult to hear important questions. If you love the smell of tar, avoid the window seats until the parking lot is resurfaced.

3. Be prepared.

Being prepared for class by skimming recommended readings, reading (yes, actually reading) required readings, and working suggested problems cuts down on anxiety and on the amount of time you spend preparing for a test or exam. Going to class with a basic understanding of a concept allows you to ask questions in or right after class, instead of having to go to a professor's office hours or hunt down the TA. Additionally, the class will be more interesting if you understand the material before coming in because you are spending less time digesting the information and comparing it to your own knowledge. The second time through the information, you are more apt to *listen* to what the professor is saying and the context in which it is being presented. The material has to be read/studied/looked over *sometime,* so just do it before class and save your time for merrier activities.

4. If you are *still* unclear about something, talk to the professor after class.

Class time isn't always the best time for a personal Q&A opportunity with a professor. Discuss the subject after class. If you have another class or your professor seems to be in a hurry, make an appointment before leaving. Ask questions as they occur so you aren't playing catch-up later.

5. Keep at it.

When you are taking a poorly taught, difficult class, motivation to do homework and show up for class, let alone pay attention, may be low. There are ways to get around it though.

You and Your Professor Are on Asynchronous Wavelengths. How to Learn Something in Poorly Taught Classes

You probably already knew that everyone learns things differently. What you probably didn't know is that engineering education specialists Richard Felder and Linda K. Silverman have delineated four dimension of learners in engineering: active/reflective, sensing/intuitive, visual/verbal, and sequential/global. They have found when a student struggles with a class, it is probably because the professor teaches in a different style to how you prefer to learn. For more information on this go to Professor Felder's webpage:

www.ncsu.edu/felder-public/ILSpage.html

continued

There you will find a 44-question quiz[1] to determine your preferred learning style and what you can do in class knowing this about yourself. (There is lots of other good stuff like tips for test taking and giving oral presentations in "Handouts for Students" on Dr. Felder's home page. A favorite is the cleverly named "An Engineering Student Survival Guide.")

SAVE THAT SYLLABUS

syl·la·bus, sil'a·bus, *n. pl.* **syl·la bus·es,** *syl·la·bi,* sil'la·hi" *An outline or other brief statement of the main points of a discourse, the subjects of a course of lectures, the contents of the curriculum.*

—*Webster's College Dictionary,* 1991

The first day of a class can be fun. It's a chance to find out who's in your class, speculate about how interesting the material and professor will be, and flip through a new textbook with a great title that intimidates your nonengineering roommates. The handing out of the syllabus signifies the end of vacation and the start of an intriguing new flood of information.

The syllabus also offers the rules of the road for a class and assistance in prioritizing your list of homework to do and subjects to study. Can I drop my lowest test score in this class? Are any tests open note or do I need to spend time memorizing the equations? How do the help sessions, office hours, tests, and problem sets in this class fit in with my other classes' help sessions, office hours, tests, and problem sets *and* my life schedule? Look over the syllabus and pick out what you think is important and note why. Exactly how you will juggle your other classes and activities while understanding this course's material can be hidden in the syllabus. Mark up the first CE 180 syllabus on the next page, then we'll compare notes (on page 62).

---

[1]Click on "Index of Learning Styles."

Mark This Syllabus Up! What's Important Here?

CE 180—Elementary Statics

Instructor: James Finlay
Office: Jacobs Hall, Rm. 620, 662-8537
E-mail: suspsn@bridge

TA: John A. Roebling
Office: Jacobs Hall, Rm. 166B (no phone)
E-mail: brooklyn@bridge

Class: MWF 8:10–9:00,
Rm. 328, Minami Hall

Textbook: *Vector Mechanics for Engineers,* 5th Ed., by Clifford Paine
Homework: Assigned on Fridays and must be turned in the following Friday
Help sessions: Wed., 3:00–5:00 P.M.
Office hrs.: By appt. only

Topics to be covered:	By:
1. Introduction	9/10
2. Forces in a Plane and Space	9/17
3. Equivalent System of Forces	9/24

TEST 1, WED., SEPT. 29

4. Equilibrium in 2-D and 3-D	10/08
5. Centroids of Lines, Areas, and Volumes	10/22

TEST 2, WED., OCT. 27

6. Analysis of Structures	11/05
7. Friction	11/12

TEST 3, WED., NOV. 17

8. Moment of Inertia of Areas and Masses	12/02
Review	

COMPREHENSIVE FINAL EXAM, WED., DEC. 16, 2 P.M.

Grading:
Homework	20%
Project	5%
Tests	45%
Final exam	30%

Honor Code:
  Applies to all tests, the exam, and the project. You may work together on
  homework assignments, but copying is not permitted.

Okay, so you've marked up the syllabus (it's nice to scrawl in a "textbook,"
isn't it?). Now turn the page to see how your notes compare to the breakdown.

In Class—More to
Staying Awake
Than Taking Notes?

What Am I In For?

(A)

CE 180—Elementary Statics

Instructor:	James Finlay	(Fin-lee)
Office:	Jacobs Hall, Rm. 620, 662-8537	
E-mail:	suspsn@bridge	← best way to contact

(B)

TA:	John A. Roebling
Office:	Jacobs Hall, Rm. 166B (no phone)
E-mail:	brooklyn@bridge

(C)

Class:	MWF 8:10–9:00,
	Rm. 328, Minami Hall

(D)

Textbook:	*Vector Mechanics for Engineers* 5th Ed. by Clifford Paine
Homework:	Assigned on Fridays and must be turned in the following Friday
Help sessions:	Wed., 3:00–5:00 P.M. TA answers questions
Office hrs.:	By appt. only ⟶ e-mail or phone first

(E)

Topics to be covered: By:

	By:
1. Introduction	9/10
2. Forces in a Plane and Space	9/17
3. Equivalent System of Forces	9/24

#s correspond to text chapters

(F)

TEST 1, WED., SEPT. 29

4. Equilibrium in 2-D and 3-D	10/08
5. Centroids of Lines, Areas, and Volumes	10/22

TEST 2, WED., OCT. 27

6. Analysis of Structures	11/05
7. Friction	11/12

TEST 3, WED., NOV. 17

8. Moment of Inertia of Areas and Masses	12/02
Review	

(G)

COMPREHENSIVE FINAL EXAM, WED., DEC. 16, 2 P.M.

(K)   (H)

Grading:

Homework	20%	⟶ 10 problem sets
Project	5%	
Tests	45%	
Final exam	30%	

(I)

Don't be late!
E-mail if missing class
Missed tests are zeroes

Honor Code:
  Applies to all tests, the exam, and the project. You may work together on
  homework assignments, but copying is not permitted.

(J)

A. Fun with phonetics.

Do you know how to pronounce your professor's name? Professor Finlay who pronounces his name Fin-lee' isn't terribly hard to remember, but pronouncing Professor Xudong Liu (Z-Dong Lew) might be a bit tough if you forget to write it down.

B. Drop-ins welcome?

Make note of how to contact your professor should you need help, want to discuss a test, or need to get a drop card signed.

C. The teaching assistant (TA).

Does the professor prefer that you take homework questions to the TA rather than ask him or her?

D. Don't buy the wrong book!

Because engineering textbooks can be so expensive, many students purchase used texts from upperclassmen or student organizations. Before buying used texts, however, make sure that the text you get is being used in the class. Different editions often rearrange topics and have different exercises and problems.

E. Help!

Do you already have a lab on Wednesdays from 2:00 to 5:00 P.M.? Talk to your professor if you have an *academic* conflict (intramurals don't count) with any offered help sessions or office hours.

F. Check the amount of material to be covered in the noted time frames.

Chapter 1 (Introduction) is only 15 pages, most of it review—no problem. Chapter 5, however, might be 70 pages and it is covered in only two weeks. Flip through your textbook to become aware of when the material will be more comprehensive and time-consuming, and adjust your own schedule and priorities accordingly.

G. Put 'em on the calendar right away!

These kinds of surprises aren't fun.

H. Note how much everything is worth.

A quick calculation from this syllabus will tell you that each homework set is worth 2 percent of your total grade. When the going gets tough and there's too much due at one time in too many classes, make your trade-offs in the classes where you'll get penalized the least.

I. Professor irks, quirks, and policies.

Most professors will tell you what they appreciate and what bothers them on the first day of class. Make note of the attendance policy and any other "rules." Some professors will request that you hand in problem sets at their offices instead of bringing them to class. Others may even request that you sit in the exact same lecture hall seat for the whole semester so that they can learn and remember *all* 113 of your names. As funny or uptight as some of these requests may seem, do honor them.

J. Am I cheating?

If the syllabus or the school does not *spell out* what constitutes cheating, ask the professor. The university's honor code can be interpreted differently by different professors. Because you were able to compare answers in one class does not mean you are permitted to do so in all your classes. In engineering classes (especially computer science) there are a lot of gray areas that must be clarified for your own protection and understanding.

Too bad the old spelling didn't stick. Sillybus *is Latin for* parchment label.

K. "Comprehensive" means everything.

No forgetting anything after the tests; everything must be *retained!* Some courses will not have cumulative exams, so enjoy these breaks and use all the extra mind space for other classes.

> *College is no place to learn engineering or complex math (you're likely to forget all those "memorized" facts, formulae, theories, and recipes in a year anyway). However, it is a great place to learn how to learn, and about yourself. With those figured out, anything is possible (including complex math, if that's what you want to do). An engineering education can sharpen your approach to solving problems creatively and coherently—precious qualities in most endeavors. Learning is a contact sport. Ask questions. Engage. Involve yourself. Embrace new ideas and cultivate them. Strive for new experiences for their own sake. Don't focus. Pay more attention to concepts than facts, learning than grades, and people than books. Whether you are building a racecar or building an organization, there will always be people involved and new concepts to conquer.*
>
> (B.T., Mechanical Engineering '97, Cornell University)

LAB PERIODS AND THE LAW OF LABS

Engineers have a lot of laboratory periods, sometimes as many as four a week during certain semesters or quarters. Lab classes enable us to make the connection between the theory taught in class and the application used in practice. Labs are an opportunity to test our brilliant abilities in creative problem solving and help determine how much we've learned in class, and, ho hum . . . yes, how well we follow directions.

Freshman year the labs will be in the introductory sciences: chemistry, physics, and maybe some computer work and/or biology. Later the lab classes will correspond to your introductory engineering and discipline-specific classes. The important thing to remember is, if you follow the directions:

The results to laboratory experiments are completely predictable (and it's really difficult to bomb them).

This is the *Law of Labs,* and it applies to all undergraduate engineering experiments. Of course, you might not look at your watch often enough and your beakered liquid crystallizes instead of merely turning pink, but that's your fault. If it's not your fault (which will happen), the TA will usually take pity, then help determine the problem, and often provide a remedy or "better" data to use for the write-up.

Although predictability makes it fairly easy to do well in these experimental labs, just how much you get out of these periods is completely up to you. Laboratory periods can be standing marathons that also test your patience, but don't let them be a waste of time.

> *Skim the lab before going.* Knowing exactly what must be accomplished for the experiment helps you recognize little problems before they become big problems, what dangers may be encountered (uh oh . . . I think I burned a hole in my sleeve), and how the experiment ties in with theory and scientific history.

- *Ace those lab quizzes!* Lab quizzes exist to ensure that you know what is going on when you come to lab. Doing well on them usually only requires you to pay attention in lecture and skim the lab manual.

- *For team experiments, have a rotating group data transcriber.* It's easy to go into a lab session, write down numbers that are called out to you, and leave the laboratory without any real idea of what happened with the experiment. On the other hand, keeping track of data for one trial gives you a head start on how well the data will match theory when you write up the report. Catching a funky number in lab, with a TA there to explain it, is preferable to thinking it is an error after you get home. Lab sessions are much more entertaining—and educational—when you get a chance to do everything.

- *Write in your lab manual.* If your lab manual is wordy, highlight key instructions. Write results, helpful hints from your TA, and your lab partner's full name, e-mail address, and phone number—anything and everything you'd rather not lose and will need when writing up the report or results when you get home. Studying for laboratory exams is made easier by flipping through the note-supplemented manual at the end of the semester instead of going through each lab to determine and memorize the theory, applications, and results.

✤ *Your TAs are the messengers.* They are students too! Most TAs are graduate students or upperclassmen majoring in the department of your laboratory. They can relate to and remember the frustration of dud experiments or spending an afternoon or weekend on a circuit design or computer program that just *won't* work. This subject is what interests them, so don't be bashful about asking questions or going to their office hours.

A Final Note . . . Maybe I'm Not Cut Out to Be an Engineer
Don't forget that introductory classes are rumored to be weed-outs (in both engineering and premed). The first few years may challenge you. Some of the required engineering classes may not even be in your major *or to your liking,* which makes studying and keeping up difficult. But don't be intimidated or frustrated by this; think of it as exercise for the mind and stamina building—the curriculum for the versatile and well-rounded engineer. After the storm, the calm (and sun!) will come.

    If you really do think your place in life is not as an engineer, make your decision to go nontech because you suffer from a general lack of interest in problem solving and technical fields. Excellent engineers and other professionals with engineering degrees were not always the best engineering students.

Don't worry about your difficulties with mathematics, I assure you that mine are still greater.

—ALBERT EINSTEIN (1879–1955)
German-American physicist, in a letter to a student

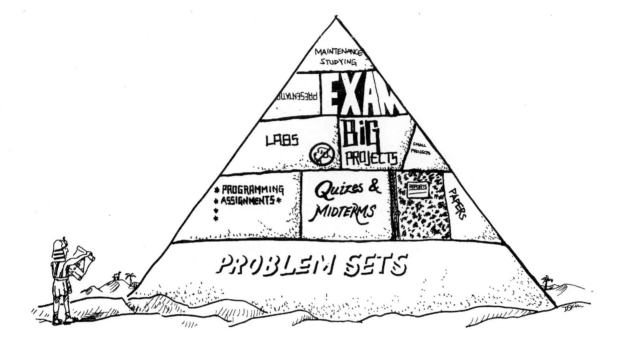

CHAPTER 6

Outside the Classroom— Workload and Studying

A person does not become clever by carrying books along.
—Proverb of unknown African origin

So you've been warned about the workload. Lots of work and unfortunately (fortunately?) there isn't much time for maintenance—you know, the day-to-day studying that reinforces the homework sets and class lectures. What makes engineering hard and so much work, of course, *is* the workload: problem sets, lab reports, and projects. There is always work to do, but the bulk of the work in engineering often hits in a tidal wave around the middle of the term. Everything is due at once. Midterms and tests seem to fall routinely in the same week. So, unless you like slam cramming, maintenance studying is important.

Maintenance studying for engineers is quite a bit different than that for our counterparts in liberal arts. Let's define maintenance studying as the time spent reading the text or reviewing class notes enough to know what is going on in class. With an English literature class, you can usually understand the lecture without having to read the book, even make a few intelligent comments (just make comments on other comments, right?). In engineering, on the other hand, you get so busy doing the assigned problems that looking at notes or skimming ahead in the text often doesn't happen. Sometimes this works out, but usually skipping on maintenance results in transcribing lecture notes without really understanding what's going on.

In this chapter we will consider the busy work and maintenance work necessary to stay on top of everything.

PROBLEM SETS

Problem sets are a way of life for engineers. It is most likely that you will complete *at least* one problem set each week of your engineering education; for some quarters/semesters, this number could climb to six or more per week.

Problem sets can be of the homemade handout genre, questions assigned from the text, or random problems your professor may have you download from the class website.

Problem sets define the busy work of engineers. While our nonengineering friends may claim they have just as much to do (reading assignments), the difference is that skipping one problem set greatly reduces an engineer's chance of pulling an A. *To do well in your engineering classes, you must do the assigned problem sets, whether the professor collects them or not.* While the frequency and relentlessness of problem sets can be stressful and tiresome, there *are* some good aspects.

The Bright Side of Problem Sets

1. They teach you what you need to know. Learning material while doing problem sets saves you from having to power-absorb the material later on before a test.
2. Doing a problem set is always better than writing a paper.
3. Problem sets make great practice problems for tests.
4. They force you to solve problems and communicate the results clearly and logically.
5. You get to know your classmates through answer-checking phone networks and late night peer help sessions.
6. You really do learn a lot.
7. You have a valid excuse to get out of annoying social engagements.

It isn't your imagination— engineers have a higher course load than all other students. Studies by the U.S. Department of Education over the past 30 years have shown that engineering students consistently have to complete notably more semester credits than any other major. On average, engineering students need 5 percent more units than physical science students (who are second), and 11 percent more than social science students (who have the lowest).

OK, you've bought a pad of the fab green engineering paper, a mechanical pencil with a cushy grip, a big eraser, and an antislip cork-bottomed metal ruler. Now let's quickly go through the elements of a problem set and a solved problem.

The Format of the Problem Set

	Assignment	Course number & title, section		Name, date	# total
	From the problem statement:				
	Given: Paraphrase the problem.				
	Find: List all the variables to be solved and questions to be answered.				
	Diagram and data: Draw a sketch of the system, labeling all relevant information from the problem statement. List the provided data needed to solve the problem. The data can be quantitative (i.e., T=74°F) or qualitative (i.e., converging-diverging nozzle).				
	Showing off your work:				
	Assumptions: Briefly state the assumptions you used to simplify and/or solve the problem. For example, if you were solving for how long it took for a water-				

melon to fall from a 2,000-foot-high building, an assumption might be: Air resistance was assumed to be negligible.

Analysis: Using the data and your assumptions—solve the problem! Start with the governing equation(s) and progress logically through the problem. Add comments that might explain choices, external information used (such as charts or tables), and how you manipulated equations. The grader should be able to logically follow your work.

Final answer: Box or double-underline the answer. Don't forget units!

Sample Problem

Hmk #2:	2.15, 2.28, 2.36 3.1, 3.6	EE 112-01 Introduction to EE	J. Wheatstone 1/27/01	1	10

Problem 2.15

Given: Figure 2-15 shows a circuit whose elements have the following values

$\mathcal{E}_1 = 3V \qquad \mathcal{E}_2 = 5V$
$R_1 = 2\Omega \qquad R_2 = 4\Omega$

Find: a) the currents in all branches
b) the potential difference between points a and b

Assumptions: Current directions shown in figure.

Analysis:

a) Using Kirchhoff's junction rule at a: $(\Sigma i_{in} = \Sigma i_{out})$
$$i_3 = i_1 + i_2 \qquad\qquad (1)$$
Using Kirchhoff's loop rule:
Left loop ccw starting at a
$$-i_1 R_1 - \mathcal{E}_1 - i_1 R_1 + \mathcal{E}_2 + i_2 R_2 = 0$$
$$-2i_1 R_1 + i_2 R_2 = \mathcal{E}_1 - \mathcal{E}_2 \qquad\qquad (2)$$
Right loop cw from a
$$+i_3 R_1 - \mathcal{E}_2 + i_3 R_1 + \mathcal{E}_2 + i_2 R_2 = 0$$
$$i_2 R_2 + 2i_3 R_1 = 0 \qquad\qquad (3)$$
Solving 3 equations for 3 unknowns:
$$(1)\ i_3 = i_1 + i_2 \quad \rightarrow \quad i_1 = i_3 - i_2$$
Plug (1) into (2):
$$-2(i_3 - i_2)R_1 - \mathcal{E}_1 + \mathcal{E}_2 + i_2 R_2 = 0$$
$$-2R_1 i_3 + (2R_1 + R_2)i_2 = \mathcal{E}_1 - \mathcal{E}_2 \qquad\qquad (4)$$
Solve for i_2 in (3) and plug into (4)
$$-2R_1(i_3) + \left(\frac{-2i_3 R_1}{R_2}\right)(2R_1 + R_2) = \mathcal{E}_1 - \mathcal{E}_2$$
Solve for i_3:
$$i_3 = \frac{(\mathcal{E}_2 - \mathcal{E}_1)R_2}{4R_1(R_1 + R_2)} = \frac{(5V - 3V)(4\Omega)}{4(2\Omega)(2 + 4)\Omega} = 0.167A \quad \rightarrow \quad \boxed{i_3 = 0.167A}$$

continued

Plug $i_3 = 0.167A$ into (2) to solve for i_2:

$$i_2 = -\frac{\mathcal{E}_2 - \mathcal{E}_1}{2(R_1 + R_2)} = -\frac{(5V - 3V)}{2(2 + 4)\Omega} = -0.167A \qquad \boxed{i_2 = -0.167A}$$

Solving (1)

$$i_1 = i_3 - i_2 = 0.167A - (-0.167A)$$
$$= 0.33A \qquad \boxed{i_1 = 0.33A}$$

b) Potential difference $= V_a - V_b$

$$V_a - i_2 R_2 - \mathcal{E}_2 = V_b$$
$$V_a - V_b = i_2 R_2 + \mathcal{E}_2$$
$$= (-0.167A)(4\Omega) + 5V$$
$$= 4.33V \qquad \boxed{V_a - V_b = 4.33V}$$

Of course, you need not follow this entire routine for a problem set in, say, calculus or physics. It is generally reserved for engineering classes where the problems present situations that require decision making based on assumptions. When time is tight, everything can be shortened and paraphrased.

Here are some things that make problem sets less of a burden:

1. *Look for a subject study guide or a problem-solver book for the tough classes.* English majors have *Cliff* (or *Coles* for the Canucks) notes, and engineers have *Schaum's*. *Schaum's Outline Series* and *REA's Problem Solvers* are great resources for both completing assignments and studying for tests. Sometimes, the exact problem you are working on is in one of these books. These books are easily found online, or at most university bookstores and libraries.

2. *Check the back of the textbook for answers.* Don't be the sad and sorry student who struggles through the entire semester before realizing the author has included answers to the end-of-the-chapter problems in an appendix.

> ⚠ **Beware!** The answers in the back of the textbook are not always correct. If you are confident you have solved a problem correctly and it checks out with other perplexed classmates, but not the book, you are probably right and the book wrong.

3. *Look for textbook example problems and problems that weren't assigned (but have solutions in the back) that might give clues to the tough problems.* Even though the problems are different, there may be a common method or a hint that isn't obvious in the assigned problem.

4. *Get old problem sets and tests from upperclassmen.* Test problems often come from the problem sets of previous years—but because of that, see the box "Old Tests and the Honor Code" on the next page.

5. *Look at problem sets early.* Not necessarily to do the problem set, but to take a quick inventory of how hard it appears to be, how much time it will take, and if there is any material that may send you to the professor's or TA's office for assistance.

☞ Old Tests and the Honor Code: The Great Student-Faculty Debate

If you seek out tests from previous years for the class you are taking right now, are you in violation of the honor code? The answer is (not surprisingly): *It depends.*

If your professor or school has a policy where this is stated specifically to be illegal—it is, of course, illegal and your using these tests as a study aid could jeopardize your college career. However, most often both honor codes and faculty do not directly address this because—what is so wrong about studying from previous years' tests? The answer is *nothing;* you are by no means cheating, you are studying. Old tests are handy to students because you get not only more practice problems but also a sense of the test style: 30 short answer, 1 long problem, or somewhere in between. But in essence, the problem is not one of cheating, but one of fairness and faculty workload. Faculty members that are generally against allowing students to use old tests as study aids claim accessibility to old tests is unfair: specific student organizations keep files to which not all students are privy. The other problem from the faculty point of view (and this is never said outright!), is that they are forced to generate new questions for every test or every exam. For a professor who has been teaching every quarter or semester for 20 years, this may become difficult without repetition.

So what does this all mean to you? Ideally, professors would be okay with students using past tests to study from and make their own tests available to everyone, but this is rarely the case. So what can you do?

- Ask your professor or e-mail an honor council rep for clarification on using old tests.
- If you do believe that some students have an unfair advantage, ask for the TA or professor to make some old tests available to *everyone* online or in a library reference folder (one or two years' worth is enough).
- Look elsewhere for practice problems: online (at other schools' old tests!), study guides, and different textbooks.
- Make tests available if you become a professor!

---

6. *Don't waste time pulling your hair out—see the TA.* If you have worked, reworked, and re-reworked and your answers still aren't checking out, then it's time to make an appointment with the TA. You don't have time to wait for a revelation.

7. *Find a classmate with whom you work well.* Two working together seems to be the optimum number for completing and understanding required assignments. Working with another engineering student also offers an additional perspective when tackling problems that require some creativity. Teaching material to someone else also ensures that you understand it.

☞ A Note on Problem Set Shortcuts

Shortcuts offer the fastest way to get from one place to another. So, the fastest way to get a problem set completed is to get a few pointers from the TA, a few hints from a friend, and some close examples from the textbook. With many bits of help, it is often possible to get the problem set finished *without* really understanding the material. Shortcuts should be used to keep yourself sane . . . and that

continued

can be OK when you are in the middle of midterms and your roommate is circulating a petition demanding that you do your laundry, *but* the central purpose of problem sets is to learn the material. As a student, however, your goal is often just to get them done, so make things easier on yourself—use problem sets to learn the material. If you're having a bad week, make time later to go back and understand what you missed. It will make for a much more benign exam week.

PROGRAMS

If you are a computer science major, your programming assignments may far outnumber your problem sets. All engineering students take at least one programming class. To many, computer science is an engaging challenge—a return to logical and creative problem solving. To others, programming is a frustrating game.

Good programmers are patient. They are persistent. They have to be. Every new programming assignment can look like a distant mountain to climb with the summit hidden in clouds. Even those who love programming say that there is a wall you hit every time. But given some time, patience, and persistence, the *Aha!* comes.

The term Computer Bug *was coined in the 1940s when an insect got into a relay in a primitive electromechanical computer, causing computer malfunctions.*

- *Make variable names obvious.* Be as descriptive as possible. Short variable names may be more pleasant to type, but they are much more difficult to keep track of when debugging your program.

- *Look for different perspectives.* Ask for different explanations of difficult concepts or problems.

- *Comment. Comment. Comment.* Commenting your code ensures that you and the grader understand what you are doing. Have fun with it.

- *RTFM.* Read the foolish manual (or something like that . . .).

- *Be elegant.* If your program works, how can you get a low grade? Think about it this way: You could build a crooked bookshelf from old splintering boards or you could build a beautiful teak bookshelf sanded to perfection. Both hold books. Which one would you want?

- *Planning puts you in touch with your inner program.* Coding without a plan can be disastrous. Before you jump into coding, sit down and plan out your structure on a piece or paper. Make an outline and write pseudo-code. Experienced programmers often use these notes as the comments in their program.

LAB REPORTS

Laboratory classes and their reports can be highly time consuming with seemingly little payoff. Although to the time-managing eye, lab reports can take way too long to prepare and are often worth fewer units than a regular class (for which you spend far less time doing outside work), lab periods and experiments are neat because you actually get to see the application of all the taught theory and scientific law. It's a chance to see how things really work instead of reading how they *should* work.

You remember the first law of labs: *The results of laboratory experiments are completely predictable.* Now consider its corollary:

If you understand the theory behind the experiments, lab classes are an easy A.

Think about it! How can you get anything less than an A when you already know the results *before* doing the experiment? After attending the lecture section of a lab and reading the directions, you could probably write the entire discussion section without the data because undergrad lab experiments always demonstrate some kind of engineering theory that is important for you to understand.

So why doesn't everyone get A's in lab classes? There are two ways to biff in lab classes:

1. Turning reports in late.
2. The write up ain't written too good.

Because lab reports *are* time sinks and often rank low priority on the assignment to do list, it's not hard to fall behind and end up handing your report in late. This isn't a big deal if you can make up labs at the end of the semester, but if you can't and points are deducted for late reports, you could be in trouble. Although it seems obvious, if you get behind on one report you will have to write two reports next week. Not a nice thought, eh? Remembering that should be enough to help you get the report in on time. However, too many students get stuck consistently handing in late reports (with points off) when they are actually working at the *same rate* as the students who are getting them in on time.

> **On the subject of tardy reports.** If it looks like you absolutely can't get a report in on time (it happens to everyone at some point), call or e-mail your TA and ask for an extension. Explain what else is going on in your life that is preventing you from completing it. Calling is better than sending e-mail because the TA can hear your pain, but note the TA sympathy factor is directly proportional to the amount of time *before* the submission deadline that you contact him or her.

Points are docked off pleasantly punctual lab reports because of careless errors: minute things that seem really obvious but are often overlooked when you are hurriedly trying to finish up before the 5:00 P.M. deadline.

> *Don't use pronouns!* No no no (unless your TA says it's OK). Also, the standard voice for engineering lab reports is passive although many professors no longer require it. Sentences such as "Then I measured the outer diameter of the spherical tank" should be "A measurement of the spherical tank's outer diameter was then taken."

- *Double-check directions.* Did you answer all the questions? At some schools, the question section of the report is the most rigorous and important part.
- *Harass the TA until you do understand.*
- *Run spell check and reread.* It's a hassle . . . you are tired, but spelling mistakes are badd, typoa worse, and misuse is not so good two.
- *Label everything.* Tables, diagrams, and especially charts and plots. Take a look at the neglected graph below. What's missing? Go ahead and write on the page.

The Unlabeled Plot

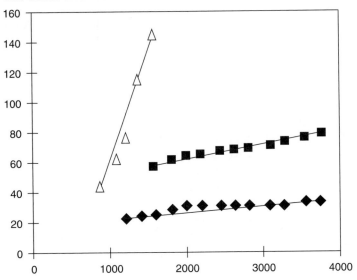

A lot is missing! The figure on the next page shows how a plot ought to look.

It's tough to remember it all, isn't it? Check out "A Quick Guide to Slick Graphs" on pages 79–80 to make sure you got it all.

The Overachiever's Lab Format

The format and length of the required lab report can vary from year to year and class to class, depending on how it was done when your professor went to college and what little things he or she thinks are most beneficial to you. Vital components of the well-thought out lab are usually covered in the first class of the semester or quarter. The following format illustrates a good way to go if your professors don't spell out what they want.

> **Use the first lab as a template.** After you write the first lab for a class, save it to disk and use it as a template for the rest of the quarter or semester. Typing over labs saves time formatting and inserting Greek symbols for variables.

0. Cover Page
Include the title of the lab, the lab number, the date submitted, the class title and course number, the names of any group members, the professor's name, the TA's name, your name, and your e-mail address.

The Labeled Plot

Graph 1. Pump Performance at Constant Head

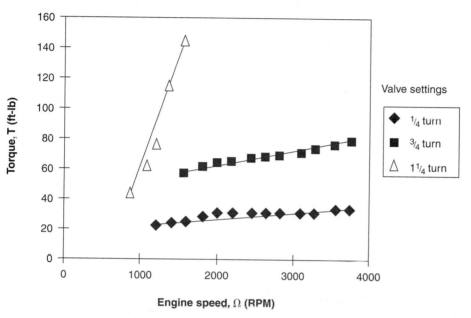

Charles Duryea 10/6/02

I. Abstract
In 250 words or less, sum up this lab experiment: the purpose and motivation of your experiment, how the experiment was carried out, your main findings, and the conclusion.

> 👉 **Who is your audience?** Abstracts (and even the whole report) can vary greatly depending on the audience for whom you are writing. References disagree about who you should assume your readers are: peers (classmates) or professionals (as if you were a practicing engineer). Ask your TA or professor, but keeping things in perspective, your best bet is to write for a single nitpicky grader who is pretending to be uninformed regarding the experiment and results.

II. Introduction
A. Statement of Purpose
This first section or statement should simply state what the purpose and motivation of the experiment is and how the experiment will be observed or studied. A statement of purpose is usually a few sentences long.

B. **Definition of Variables**

Define all the variables used in calculations and discussion. Variables are listed in alphabetical order with Greek symbols first. Appendix D lists the Greek alphabet so you can even get those in alphabetical order!

> **Do the definition of variables section last.** With more than five variables, it can be more time efficient to wait to do this section after keeping a running list of the variables *as you use them* in the write-up and calculations. Doing this will save time that otherwise would be spent flipping back and forth through your report to add the ones you missed.
>
> **Rearrange your tool bars.** Set up your word processor or spreadsheet tool bars to include frequent use buttons (like the equation editor, symbol, superscript, and subscript buttons) that aren't already there.

III. **Background**

A. **Equipment Setup**

List the main equipment (rulers, pencils, and so forth can be assumed). Give enough detail so the reader has an idea how to reproduce the experiment. If an equipment list is already in your lab manual, just copy it. The following are acceptable:

Anemometer
DC Power Supply, 0–40V

These, of course, are better:

Mini Anemometer, Kurtz series 490, Model 490, 0–200 or 0–2000 SFPM, ±1.25%
DC Power Supply, HP LVR series, Model 6266B, 0–40V, 0–5A, ±0.1%

Slightly geeky, indeed, but there is a reason. If later researchers wanted to duplicate your results, they wouldn't want to use any old anemometer or power supply, but one as similar as possible to what you used in order to minimize deviations for controlled comparisons.

B. **Experimental Method**

This should already be in your lab manual. All you have to do is convert it to your own words and the passive voice. If the experimental method isn't in your manual, remember to take good notes during the experiment. The easiest and clearest way to express the method is in numbered steps.

C. **Theory**

What are the governing equations that describe this experiment? Again, these should be in your lab manual; if not, just pull them from a textbook.

> ⏱ **Number your equations.** Number all the useful equations you mention in your lab. Just like in textbooks, numbering makes it easy to refer to the equations in the discussion section instead of typing out the proper name.
>
> **When time is tight, write equations in by hand.** Although both Microsoft Word and Word Perfect have good equation editors, they can be slow to use. When you are running close to the wire, consider leaving a few blank lines and writing the equations in by hand.

IV. Results and Calculations
A. Results

Results can be quantitative and/or qualitative. If you have a whole truckload of raw quantitative data, it should be placed in an appendix that is referenced from the results section. A moderate amount of quantitative data can be included in a table in this section. *However, what TAs and professors most want to see in this section are some slick graphs that show how all the data correlate to theory.* A graph says a thousand words, raw data don't say much of anything. Check out "A Quick Guide to Slick Graphs" to make just that.

A Quick Guide to Slick Graphs

Another engineer should be able to pick up a slick graph and be able to completely understand what is being shown without knowing anything about the lab experiment. Let's go to our labeled Graph 1 to look at the main points of slick graphing.

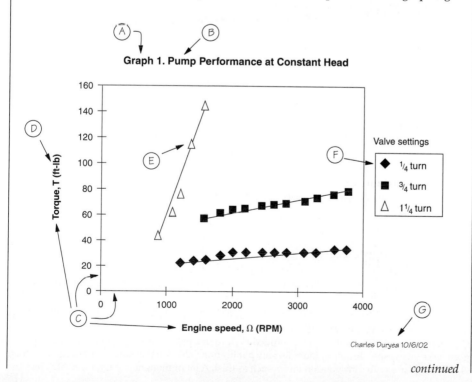

continued

A. **Label**

Labeling, like numbering your equations, makes for easy reference and a polished report.

B. **Title**

The title should briefly say what was going on that produced the results displayed by the graph. Don't restate the axes! "Torque vs. Engine Speed" sounds like a wrestling match, not a graph title. Capital letters should occur as with any title. Although the title is shown here above the chart, some professors prefer titles and labels to be below the chart.[1]

C. **Variables**

The independent variable is usually plotted on the *abscissa* or horizontal axis, while the dependent variable is plotted on the *ordinate* or vertical axis. Thus, torque is on the ordinate because it was the result of our fiddling with the engine speed. When dealing with nondimensional or dimensional quantities, the axes should be consistent; that is, make both axes nondimensional or both dimensional. Label the axis with the variable and abbreviations or symbol (if there is one).

D. **Variable units**

Don't forget them! If you have a quantity that is dimensionless, label it so with "dimensionless."

E. **Points, lines, curves, etc.**

If you have experimental data, the data points must be shown. While Mardi Gras point colors look festive on the computer screen, it is better to convert the default multicolored boxes over to symbols if you are printing on a black-and-white printer. Once printed, it will be much easier to differentiate groupings and see the overlap.

Lines and curves through data are usually "best fits" (not connect-the-dots) with a ruler, french curve, or "Trendline" if you use Microsoft Excel. Best fits are based on correlation by using a regression formula (easiest with a spreadsheet). Do use an appropriate best fit; while a 9th order polynomial fit might go through all the points, it would not represent the theory at all!

The best-fit line or curve through data is a solid line. Dashed lines are usually saved for extrapolation (when the line is extended beyond the collected data). When you have multiple lines for comparative data, this standard is ignored so you are able to differentiate lines.

F. **Legend or callouts**

Labels to differentiate data are necessary with more than one set. A legend off to the side or callouts that label the data on the graph can be used. The more concise the label, the better.

G. **Signature and date**

Some professors ask students to sign and date all of the graphs they produce and submit.

B. **Sample Calculations**

Include an example of each calculation you performed in processing your results. All the equations should be referenced or follow

---

[1]In case you *really* wanted to know the standard practice for lab report graphs, it is as follows: if figures (such as charts, graphs, or diagrams) are integrated into text, the title should go *below* the figure. On the other hand, if a graph or plot is printed to fit to one page (filling the page), the title typically goes *above* the graph. Technical journals and publications also vary in standard and give authors "kits" (templates with directions on topics such as this). A lot of thought just for a silly title, huh?

directly from the background theory section. Include any assumptions you made.

C. **Error or Uncertainty Analysis**

"Error? I just followed the directions. . . . " Error is the difference between the measured value and the accepted or "true value" given by theory. For experimental analysis in lab reports, error is usually represented numerically as a percentage or as a magnitude ± (plus or minus) the measured value. Error can also be represented graphically with range markers shown with the graphed data.

Error using inadequate data is much worse than those using no data at all.

—CHARLES BABBAGE (1792–1871)
English inventor/mathematician

V. **Discussion**

A. **Discussion of Results**

State what you learned—or were supposed to learn—from your results. Make statements that describe correlations and reference a graph or plot in the Results section: "From Graph (1), it can be seen that output torque increases with input power." Remember the Law of Labs—don't worry if your discussion section is short—all you are doing is stating experimental correlations that prove well-known theory.

☞ **Can you guess the most common grammatical mistake made by engineers (and nonengineers)?** *Hint:* it's a favorite word of engineers. The word *data*. That's right! Data is plural—or rather data *are* plural. The singular form of data is *datum*.

B. **Error**

After calculating error or uncertainty to your heart's content (maybe more), you now get to *discuss* it. Since it is unlikely that the only errors in the experiment were due to instrument inexactness, you need to mention possible sources of added error that may have affected your results. One sentence explaining which error(s) occurred and how is usually sufficient. Engineers have distinguished several types of errors on which we can place blame for our imperfection:

1. **Random.**

No idea how it messed up. Although you have no idea what happened, speculate.

2. **Systematic.**

It consistently messed up about the same every time. Example: Your watch is 10 minutes fast, so every time you make a time check, it is +10 minutes off.

3. **Human.**

You messed up. Example: You misread the ruler.

4. **Computation.**
   The device or instrument used to crunch numbers messed up. Example: You are calculating the distance to the fourth nearest galaxy with 327,682,357,883,567 significant digits and your computer can handle only 256. This error is highly unlikely for our labs.

   If your data and results are all over the place, the best thing to do is discuss the expected results for the experiment and then, with the aid of our four types of error, surmise what happened.

C. **Questions from Lab Manual**
   Some professors will have you put them here. Others will have you put them in an appendix.

VI. **Conclusion**
   Based on the results, state what was proven. The conclusion should answer all the questions that came to mind from reading the experiment's purpose.

 "In conclusion . . . " isn't a good way to start the conclusion section. But you already knew that.

VII. **Appendices, Bibliography**
   Appendixes can include raw data, error calculations, answers to assigned questions, tearouts from your lab book, and extra diagrams or plots. Bibliographies are the same as for papers in any discipline:

   Author's last name, first name. *Book/Paper title. City:* Publisher, ©.

   Example:

   Donaldson, Krista. *The Engineering Student Survival Guide,* 3rd ed. Burr Ridge, IL: McGraw-Hill, 2004.

So there you have it—the most complete lab report possible. Depending on the amount of detail required by your professor, lab reports can range from a few pages of data and comments to 30 or more pages. It is highly unlikely that your professor will require you to go through all the steps outlined above, but if he or she does, you will be in good shape.

 A final note on labs. Don't get caught up in little details. Producing a good lab report is already a lot of work, so don't get bogged down fiddling with colored plots, scanned-in photographs, overcomplicated CAD[2] drawings, and so on.

---

[2]CAD (Computer Aided Design) refers to any drawings, models, or renderings done on a computer, typically with software specifically for that purpose.

Knowing is not enough; we must apply.
—JOHANN VON GOETHE (1749–1832)
German writer and theorist

Engineering projects pop up in any course where there may be some engineering design going on (almost everywhere). Whether you choose door 1 (a heat exchanger!) or door 2 (you, too, could design your very *own* voltmeter!), recognize that projects always take more time than is budgeted. The key to painless projects is *start early*. Not a surprise, eh? We will subdivide projects into two categories: SPPs and BTUs. SPPs (small, piddly projects) and BTUs (big term undertakings) can be individual or group projects.

Small, Piddly Projects (SPPs)

SPPs are those projects that professors decide are fun ways for you to apply what you've learned in their courses. You often have only 2 weeks (or less) toward the end of the term to wrack your brain, then write your work up nicely, and maybe even present it to the class.

1. *First things first.* Understand the project assignment. Many times the problem statement is much more confusing than the problem itself.
2. *Research.* SPPs usually do not require much outside research, unless paying attention in class isn't your forte. Are there any back-of-the-textbook computer programs you need to learn? Depending on the project, sometimes a quick trip to the library or a 15-minute Web search can help substantially. Once you figure out what you need to do, what equations you'll need, what methods work best, and what kind of results you should expect, you are set to . . .
3. *Work it.* Allow yourself lots of time (goes with starting early) for computations and problem solving. Plug through it. If you need to write a computer program, don't forget to allow time for debugging and sanity breaks.
4. *Test it.* Does everything work without glitches? Your work should be easy for the professor to follow when he or she grades it. Did you state your assumptions? Sometimes decisions that are clearly evident to us are not obvious to others tackling the same design problem. Look critically at your project for any potential holes that could cost you.
5. *Fix things.* Rework it. Add. Subtract. When running short on time, fill in as much as possible and list what you might have done had you not run out of time.
6. *Write it up.* You are almost finished!
7. *Go the extra mile.* It takes very little time to polish up a project. Throw all those sources into a bibliography. Draw a diagram that clarifies final concepts. Whip out a spreadsheet graph. Number your pages.

Big Term Undertakings (BTUs)

And you thought BTUs were British thermal units! You just can't unknowingly and accidentally find yourself in a class with a term project (a.k.a. big undertaking). A BTU is not an SPP with more time to think. You should have been forewarned by upperclassmen and even by professors. Big term undertakings are often the finale—the end of a course series or maybe the end of your undergrad career (senior projects!). They tend to reflect real-life engineering and project management. Although BTUs are a *lot* of work, it is cool to see the engineering theory you've learned come together to produce something impressive. A good term project will make you an expert in your chosen topic.

1. *First things first.* Understand the project assignment and what is expected. Master the acronyms. Learn (or relearn) how to do literature searches at the library, operate machine-shop tools, use computer codes you have almost forgotten, and so forth.

Engineers developing the Nintendo 64 consumed 5,000 bagels during the project (and a lot of cream cheese, too).

> **Eeny meeny miny mo.** If you are able to pick your own term project topic, decide as early as possible. Choose something you really like—besides consuming your life, it could maybe even lead to a job or graduate work.

2. *Research.* It is this phase of the BTU that consumes the largest chunk of time; as much as half of the term can be spent getting up to speed on the assigned or chosen topic. Any good project that really puts you to work will cause anxiety and frustration early on. The whitecapped swell of frustration follows the realization-of-the-magnitude-of-what-you-have-gotten-yourself-into wave that washes over you as you weed through all the background information. *First,* start with the easy stuff. If you are doing a project on contaminant tracking in rivers, begin with some water quality and contaminant readings from an environmental engineering text. After you get that nailed down, move on to the journal articles. Don't be afraid to ask questions of professors, graduate students, friends of distant relatives, and even people in industry.

3. *Generate.* Generate as many ideas, solutions, methods, possibilities, and paths as possible. This is a great time for a mind map (see "The Marvelous Mind Map" on page 87).

4. *Work it.* Working it means decision making by you and perhaps a computer. After the time and toil devoted to research, this step is almost disappointingly easy.

5. *Double-check results.* Does everything make sense? Check your assumptions. Double-check your constants and anything that involves a unit conversion. The answers you arrive at are often obviously good or bad for BTUs. If you find everything looks good, it's time to feel *relief*. If things look monstrous, make an appointment with the professor—there may still be time to recover. You are approaching the home stretch.

6. *Tend to the small details.* With big projects there are often many minor considerations that get tossed aside. Dot your i's, cross your t's. Pro-

duce a schematic of your solution and process. Go back and fill in the gaps.

7. *Write it up.* After all the time you have devoted to this project, don't slack off now! Writing up the BTU will take more time, patience, and disk space than expected. If possible, get a good night's sleep before the final edit. It is painful to see a typo in the freshly bound copy.

8. *Go the extra mile.* Polish it up. Scan in pertinent photographs. "Borrow" some slick graphics from the Web (it's legal for educational purposes as long as you credit the source!). Generate some CAD drawings and renderings. Put a cool graphic on the cover page.

9. *Take it to the printer for binding and an extra copy.* Don't do this at the last minute! ("The earliest we can have this done for you is tomorrow afternoon.") Pick a cool color for the cover (Kinko's has Rocket Red for the Aero/Astros). Make an extra copy. Finish out the semester with a bang.

Looking for a different kind of group project in engineering? Check out page 148.

Looking for a different kind of group project in engineering? Check out page 148.

Group Projects

After a couple of years of being almost entirely self-reliant in the completion of problem sets, assignments, and labs, engineering students get thrown into groups. ("You mean I *have* to work with someone else?") Group projects often make their first appearance the junior or senior year in design classes, but at more hip schools they are assigned as early as the freshman year or for laboratory classes. Although it may feel completely unnatural to work with other people, engineers in industry almost always work as part of a team. Who could design and build a 727 aircraft single-handedly? Group projects add a whole new dimension to working on a project; everyone has different opinions, different schedules, different expectations, and different amounts of time and energy they are willing to commit. How should you tackle it?

1. *Choose a group leader.* Anarchy isn't (usually) a good thing. Flatter a responsible peer and nominate him or her. If you are interested in heading up your team, skip to number 8 and then come back. It is also a good idea to appoint a scribe at each meeting.

2. *Establish a timeline or—for larger projects—plan out your project on a Gantt chart.* The earlier the better. Keep it updated so that every group member is accountable and on schedule. For more about a Gantt chart, read the boxed material "Mr. Gantt Would Be So Pleased" next.

Mr. Gantt Would Be So Pleased

A Gantt chart is a schedule represented in bar-graph form used to plan the sequence of project tasks over a known time period to ensure that all deadlines and important milestones are met. Tasks are listed down the vertical axis in sequential order, while time is listed in blocks across the top. According to engineering lore, Henry Gantt was able to boost World War I factory production because workers related better to a pictorial representation of project progress and goals.

continued

For our wee project of building a dorm bedroom loft, the time scale is days. For most course projects, the time scale will be in weeks. Bar charts show the amount of time a task requires by spanning from the start date to the planned finish date. A dotted line with a diamond (. . . .◊) extended from a bar is the float (or extra) time (A) there may be until the task must be completed (B); that is, you want to have your wood purchased by Tuesday, but to get help from your visiting sister (a carpenter) it really needs to be purchased by Wednesday afternoon (B). Once you have started working on tasks, you can go back and update your chart by filling in an equivalent percentage of the block (C). The vertical dashed line represents the current day (D). Dependencies (E) are noted with arrows (you can't buy your wood before doing calculations to figure out what exactly you need) and milestones (F) are noted with a point.

The Gantt Chart for the Loft of Love

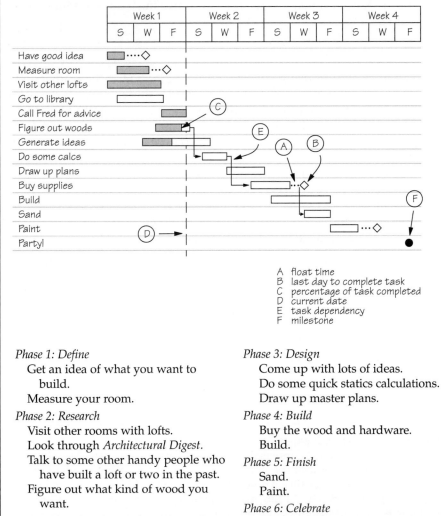

A float time
B last day to complete task
C percentage of task completed
D current date
E task dependency
F milestone

Phase 1: Define
  Get an idea of what you want to
    build.
  Measure your room.

Phase 2: Research
  Visit other rooms with lofts.
  Look through *Architectural Digest*.
  Talk to some other handy people who
    have built a loft or two in the past.
  Figure out what kind of wood you
    want.

Phase 3: Design
  Come up with lots of ideas.
  Do some quick statics calculations.
  Draw up master plans.

Phase 4: Build
  Buy the wood and hardware.
  Build.

Phase 5: Finish
  Sand.
  Paint.

Phase 6: Celebrate

3. *The mantra for brainstorming sessions is: Every idea is a good idea.* Remember that when you want to bang a group member on the head with your binder. Write down every suggestion. Encourage wild ideas—the best thoughts can come from the most random inspirations.

The Marvelous Mind Map

Mind maps are an excellent tool for recording all the many directions and ideas of brainstorms. They can be used for individual or group creativity. Although the format can take any form you wish, the mind map usually starts with the central idea in the middle of the page (or flipchart) with other ideas and thoughts radiating outward. Mind maps are great for seeing recurring ideas and patterns.

　　Want an example? Keep reading!

---

4. *Divide up tasks.* Define and divide central and secondary tasks among your group members. Everyone should have one area for which they are responsible but also have multiple secondary tasks that assist other group members. Work with your team so that all members are matched with work they like.

5. *Keep a log book.* Write everything in it and don't lose it! Why? See "What Good Is a Log Book?"

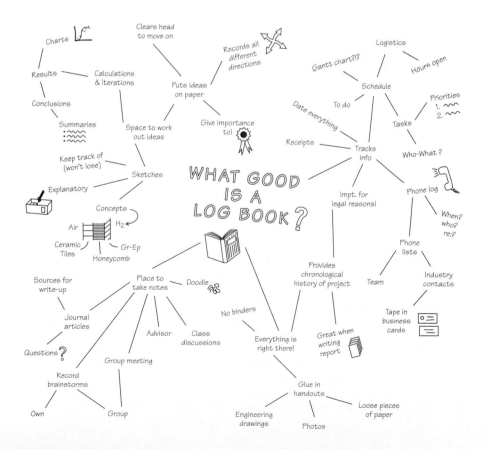

6. *During the research stage, use the first 10 to 15 minutes of each group meeting for a "round table."* The round table format allows everyone to explain briefly what they have done (battles fought and won) since the last meeting and any stumbling blocks (dragons) they may have hit. Many times one group member can help out or solve another's problem.

7. *Be a nice person to work with.* Pull your share. Volunteer to go first. Know when to shut up. Keep a sense of humor. Ask for input from any quiet group members.

8. *Recognize that the leadership position comes with perks and headaches.* If you are the type of person who has *the* vision for the final product—Go for it! Team leadership is great experience, but the person in charge often ends up doing the most work.

The larger the group the less the number of regular contributors to group discussions! Research has shown with teams of 3–6 people, everyone speaks; 7–10 most people speak, quiet people may say less or nothing at all; 11–18 people, 5–6 people speak a lot, 3–4 others occasionally; 19–30 people, only 3–4 people contribute regularly; and for 30+ people, little participation is possible.

Forming, Storming, Norming, and Performing: The Life Cycle of a Team

From studying over 50 small groups, psychologist Bruce Tuckman determined that teams move through four relatively predictable stages of group dynamics: forming, storming, norming, and performing. He found that it is not unusual for a team to get stuck in one of the stages—typically forming or storming—not a surprise to anyone who has had a bad team project experience.

1. **Forming: What? Who? When?** Your team is just getting used to each other and starting to feel like members of the group. If you have a team leader, there is a fairly high dependency on him or her to get the ball rolling. A team member or two may even change his/her authority.

   Typical characteristics of forming are:
   • Initial, tentative attachment to the team
   • Hesitant participation
   • Abstract discussions of issues peripherally related to task (possibly to the impatience of other team members)
   • Anxiety about the new situation and/or task
   • Determination of acceptable group behavior
   • Not much work gets done

2. **Storming: AGHHHHH!** This is the most difficult stage for the team. The amount of work that needs to get done sets in and team members panic. How on earth will we ever accomplish these goals in the required time? Members may become hostile, testy, or overzealous. Some will be impatient about the lack of progress, but the team as a whole is still learning and processing the problem.

   Typical characteristics of storming are:
   • Irritation with and arguing among team members
   • Emergence of a hierarchy within the team, disunity
   • Establishment of unachievable goals
   • Not much work gets done

3. **Norming: Everyone mellows out.** Folks get used to each other, accepting the team structure and rules. Members are starting to carve out their own niches and the emotional conflicts that occurred in the storming phase are replaced with more cooperative relationships.

   Typical characteristics of norming are:
   • Group effort to keep things harmonious
   • Team members may help each other with various tasks

- Everyone gets to know each other better on a personal level
- A moderate amount of work gets done

4. **Performing: Let's get busy!** Your team is now an effective, cohesive unit. Problems are quickly recognized and navigated. Team members recognize each other's strengths and weaknesses. Your team is productive (finally, you might say)!

Typical characteristics of performing are:
- Satisfaction with the team's progress
- Team members feel like they have gone through some constructive process of self-change
- A whole lot of work is accomplished here!

Teams will go through ups and downs, spirals and zigzags of emotions. In the downspins, the best thing to do is to recognize that your team is normal and things will progress with effort.

Iowa State's Get a Grip Youth Leadership Program has a great document on team building with more detailed descriptions of the phases as well as tips for encouraging constructive team work, strategic planning, strength and weakness evaluation, and icebreakers for when a hokey game is *just* what your team needs:

www.iastate.edu/~getagrip/pdfs/ch2.pdf

Presentations

If you have to give an oral presentation, don't even consider doing it without overhead projectors or slides. First of all, talks are much easier when you have overheads with which to share the audience's attention. Second, good over-heads or slides (especially color ones) make for a polished presentation that will be sure to earn you brownie points. Now it is easy to simply crank out slides in PowerPoint and hook up your computer to a projection unit.

If your professor has a specific format or outline he or she expects you to follow, simply fill in the blanks and use it. However, if freelancing, consider some of the following presentation basics:

 So many options! Presentations with visuals can take any one (or several) of the following mediums: slides for slide projectors, overheads (also called slides), computer hook-up to a projection unit, and at some schools the overhead cam.[3] Pick one that best supports the material you are presenting.

⚠ **Have a backup plan for burned out lightbulbs, missing adapters, and fried floppies.** You want your audience to remember your talk—not "how well you handled it."

---

[3]Overhead cam presentations are often possible in classrooms wired for televised recording. The speaker has a set of slides printed on regular paper sitting on top of a desk. The overhead cam focuses on the top page and projects it up on a large screen behind the speaker.

Put different topics on different slides. The rule of thumb is: **One *text* slide per minute.** If you end up with 13 text slides for a 5-minute talk, you have too many topics and need to reorganize!

 Storyboard first. Before you sit down at the computer, work out on paper what information and diagrams you want on each slide. Doing this actually saves time. Sticky pads allow you to arrange, rearrange, and rearrange topics several more times.

 Slides should highlight the main points (not be a transcript) of the speaker's talk. Details and minor points should be discussed by the speaker instead of crammed onto the slide. The fewer the words on the slide, the better.

 Slide titles and bulleted text should have the same sentence structure. It flows. It's cohesive.

 A blank bright white screen mesmerizes audiences and clues them out. When you're not using it, turn off the overhead projector or cover up the display with paper. The audience should be listening to you.

 If you have many graphs or pictures, consider running two overhead or slide projectors simultaneously. This is particularly effective when you have comparative graphs or data, or a graphic that best accompanies a text slide.

☞ **On Your Own? Here's the . . . TYPICAL TECH TALK TOOL KIT**
    It can be used for anything from a progress report to a final project presentation.

- *Title page.* Name, rank, serial number. The beginning of the talk is a good place for an anecdote that relates directly to your project/presentation area.
- *Overview of the talk.* The overview is generally a list in bullet format of the slide topics to follow. People like to know what is coming up.
- *Introduction.* Define the problem, explain the motivation, briefly discuss the experiment, and state the goal(s).
- *Background.* What background information is pertinent to an understanding of this discussion? Include the main governing equation or central theory and assumption. If appropriate, state any hypothesis based on your understanding of the background material.
- *Progress/Experiment.* For progress reports you will want to discuss what you (or your team) have contributed. For more research-oriented talks, this is highlighting the main points of your experiment or study. You want to explain how you collected your data or carried out your study. If you did traffic flow analysis for a transportation class, how did you do it? Sampling with web cams for one week? Sitting on the corner of an intersection during rush hour on one weekday and one holiday?
- *Accomplishments/Results.* Show any pertinent charts, tables, and so forth. This is the presentation equivalent to the results and discussion sections of the lab report: expected/unexpected results, findings, sources of error, etc.
- *Conclusions.* What was discovered? What do your findings mean in the big scheme of things?

- *Questions left unanswered, areas for further research.* You really would have liked to further investigate some of these other areas, but time ran out. This is the place to mention stumbling blocks and future directions.
- *Acknowledgments.* Thanks to . . . and can I take some questions?

✿ *Rehearse!* It makes such a difference—to you and your audience. Great presentations have been ruined by poor speaking or fumbling to get slides in the right order. If you tend to get nervous giving oral presentations, try to think of your talk as an explanatory discussion with friends rather than a scripted narrative. Pretend you are a tour guide and let your slides cue you.

✿ *For formal presentations, distribute handouts with copies of your slides.* PowerPoint allows you to print multiple reductions of slides on a single page in the print menu. Any additional supporting material that, because of time constraints, did not make your presentation can also be included in the handouts.

My senior year I took a biomechanics course. I was very excited about the course, and was looking forward to learning about the forces involved with the body, how muscles and bones interacted and things like this. Unfortunately, although the course description mentioned many exciting topics, it quickly became clear that the professor was mostly interested in the fluid dynamics of blood flow. And not much else. Lectures degenerated into painful slogging through equations and discussions of theory that was only partially understood by the class. Our midterm was a take home test, designed by the prof to take two hours. It took us all twenty-four, and still nobody thought they understood it. The course continued on this way all term. Class attendance dwindled. The professor handed back NO assignments, so although we were all fairly certain that we were doing poorly, no one knew for sure. It was very painful. I was actually reading the biomechanics book sections that interested me, not the ones that were assigned.

Finally at the end of the course, the professor assigned a project, requesting that each of us give a talk on a biomechanics topic that interested us. I ended up giving a talk on the mathematical models behind bipedal motion, which didn't have a whole lot to do with the stuff we had been studying, but I was pretty into it. I made a bunch of drawings of mechanical models on a big art pad I had, and on the day of the talk used the pad as an easel to talk from (before the days of PowerPoint). The professor was absolutely ecstatic! It turns out I was the only one to use visual aids during my talk, and somehow he had never thought of this idea in talks before. He just couldn't get over it! I was also one of the few who actually picked a topic that they liked, and was excited about it. I ended up getting an A in the course, in spite of the fact that I am sure I failed all the exams, along with the rest of the class. (I don't actually know, since no tests were ever returned.) The professor never failed to stop and chat with me in the hall and was always very interested in what I was doing after that.

I think that the enthusiasm and pictures during that talk made a big difference to the professor. It turned out that he actually knew a fair bit about the things I was interested in, and I learned some cool things from him. Somehow he just needed to see a student getting fired up about something before he could get on track. I think lots of people are like this. If you go and talk with them about the things you are interested in, they may actually have good insights. I'm still not sure what was revolutionary about an easel full of pictures though.

(K.H.D., Biomedical Engineering '94, Brown University)

PAPERS

Papers, optimistically believed to be rare in engineering education, are saved from extinction by the writing requirement. You may choose to fulfill the requirement with Post-Modern Ideals in Literature or Technical Writing. Whichever class you choose, pick what you like and will find useful, not what you think is easy. Writing on topics that don't interest you is not fun or easy. Not many of us will graduate and be granted a secretary as part of our first job package. Engineers in industry might find themselves devoting more time to business letters, memos, and reports than to computational work. Grad school doesn't let you escape either; there are many essays and proposals to write.

Freshman and sophomore engineering papers are lab reports; memos to fictitious companies, clients, or agencies; article summaries; and maybe an explanatory paper that accompanies an experiment or computer program. As an upperclassman, you may be assigned research papers that are rather like wordy lab reports under the guise of a paper.

We won't get too much into paper writing here other than to offer some excellent websites that cover everything from planning how to write a paper to your final edit.

- **Purdue's Online Writing Laboratory (OWL):** owl.english.purdue.edu

   Scroll down to "handouts and materials" where you will find a plethora of highly useful material, including writing tips for "English as a second language" writers. This site is so popular that it gets over 20 million hits per year—mostly from non-Purdue people!

- **Professor Darling's Guide to Writing a Research Paper:** www.ccc.commnet.edu/mla/index.shtml

   This well-organized site offers comprehensive descriptions for citations based on the style.

- **The University of Toronto Engineering Communication Centre's Online Handbook: http://www.ecf.utoronto.ca/~writing/ handbook.html**

   This useful website is tailored particularly to engineering students and covers four main topics: types of reports (such as memos, case studies, progress reports, etc), documentation (IEEE style, which is the most commonly used style in engineering publications), oral communication (presentations and visuals), and the writing process, including some good tips from graders.

- **Scientific and Technical Writing for Engineering and Social Science Students: writing.eng.vt.edu**

   This excellent website was put together jointly by four American universities and covers a range of technical work that will get you from freshman year memos through a PhD dissertation and everything in between. This comprehensive site also has writing exercises, links to places such as the "Grammar Gym," and formatting information for visuals such as charts and tables.

What is written without effort is in general read without pleasure.
—SAMUEL JOHNSON (1709–1784)
one of the first authors of a comprehensive English dictionary

MAINTENANCE STUDYING

What is there to study in engineering? Everything important is learned through problem sets, right? Not really. Only the practical knowledge is learned in problem sets, but really understanding cause and effect and the why and how comes from reading (yikes!) the text, looking over notes and problem sets as you get them back, and paying attention in class (or some combination of the above). Studying to understand what is going on in class is maintenance studying. Maintenance studying differs from test-tomorrow studying and how-do-I-do-this-homework-problem studying because it takes a higher level of motivation. Studying to learn and to understand happens for the truly inspired, but it can be difficult for the rest of us. Studying to keep up, though, never takes as long or is as painful as we might think it is.

Studying from the Textbook

Of course the shapes of these matrices must be properly matched—B and C have the same shape, so they can be added, and A and D are the right size for pre-multiplication and postmultiplication. The proof of this law is too boring for words.

—GILBERT STRANG (1936–)
American mathematician and professor,
from *Linear Algebra and Its Applications*

Engineering professors do not seem to depend as heavily on textbooks or outside readings as other instructors often do. Some professors will actually assign sections of the textbook to read; others will never mention the text at all (leaving you frustrated for wasting $80). For the most part, engineering textbooks are fairly good because they are fun to dust off later and reread (as references, of course), but more important, they provide the background and theory to pull seemingly random problems and concepts together.

So you have purchased, borrowed, or stolen the course text. What should you do with it?

1. *Read it.* Peruse it. Skim for boldface print. Even when the professor doesn't assign sections to read, use the index or table of contents to find the most pertinent sections. Once you understand the reasoning behind one concept, you will see similarities in other concepts.
2. *Learn important terminology.* Understand the professor. Impress classmates. Test it out at social functions.
3. *Work through example problems.* Homework and test problems will be much easier—*guaranteed.* Also try working through derivations without looking. When the text says: "left as an exercise for the reader," do it! No one can teach you as well as you can teach yourself.
4. *Interact with it.* If this is a text you plan to keep, don't be afraid to mark it up. Make studying from a textbook proactive: Write comments in the margins, underline or highlight key phrases or terms, tab pages with useful charts and tables, and keep a list of questions to ask the TA. Proactive studying will keep you focused.
5. *Take notes from it.* Some folks find that to *really* absorb what they read, they need to take notes on the readings. However, time will not be available to take notes on all the sections you cover in class unless you are superhuman or need hobbies. However, taking notes from book material is extremely useful if you outline sections that have been difficult to understand or appear to be very important to the course.

    Read the section first and then go back and take notes. Taking notes as you read makes it more difficult to distinguish what material is important, and you will probably write down much more than is necessary.

    Do not redraw charts or diagrams; try to translate them into your own words. You'll find that you will remember them better.

Write your book notes on one side of loose-leaf paper. With everything on one side, you can spread your notes out, and rearrange and even wallpaper without fear of missing material.

What to Do When the Textbook Stinks!

6. *Use it as a hot mat.* Perfect size with great thermal insulative properties. Unfortunately one needs only so many hot mats. Hmm . . . nothing more discouraging than a bad textbook. It's not hard to spot one because it often displays these characteristics: useless examples, not enough (helpful) examples, poor explanations of key concepts, too many words, and not enough pictures. So what do you do?

 Look for a study guide. Schaum's to the rescue again (see the section "Problem Sets" on page 72).

 Find another textbook in the same subject by a different author. Your school library is likely to have many texts on the same subject.

 Ask an upperclassman for his or her old notes. Offer same-day copy and return service.

 Go talk to the TAs. They might recommend better textbooks or prepare a "translated" handout.

 Do a Web search. You'll find that professors at some other schools put their class notes on the Web.

 Investigate the science encyclopedias at the library.

Studying from Class Notes

Why spend the energy taking nice, color-coded organized notes if you're never going to look at them again? The best time to review notes is right after the class—to reinforce concepts. A professor may occasionally cover material or give an example that is not in the textbook (a good clue to what he or she thinks is important). Use your notes! Look them over. Compile and process. Class lectures highlight the most important concepts of all the material.

Studying from Problem Sets

A problem set with an explanatory solution set is a gold mine. Homework sets serve to condense and summarize the class material (which has already refined the textbook material) down to the really important stuff. You will read about the 80-20 Rule in Chapter 7. The important 20 percent will make a debut in the problem sets before it makes an encore appearance on the test (and if it's really special it makes a comeback on the final). The point is this: Keep all homework solutions. Look over them when they are made available (don't bury them in a folder). See where you went wrong. Revel in where you were cleverly right. Notice better or alternative methods. Finally, if the TA or grader needs to readjust your homework grade, he or she will take more kindly to an immediate plea than to the end-of-the-term petition when you need two points somehow to get a B+.

☞ **Advice from the experts: How to make *time* fill the *work* allotted!**
Time management guides are available in a dizzying array and amount in print on the Web—who has time to select one and actually read it? Congratulate yourself for saving time by considering these tips from the gurus:

- *Figure out when you think and work best.* Are you a morning person or a night owl? Go with your flow. If you know that your mind is mustard after 10 P.M., tackle the "thinking work" while you have your cells intact.
- *Do the hard stuff first.* Yes, yes—this is not human tendency, but tackling the challenging concepts and assignments while you are alert is far more efficient.
- *Distinguish between important and urgent.* (It's important.)
- *Schedule yourself.* Not sure how you can possibly get everything done before Thanksgiving break? Sit down with a calendar and book yourself up—don't forget personal priorities like laundry and exercise!
- *Be a time-nazi.* Give yourself a specific amount of time to finish a task and stick with it. Even if you don't finish the task, you should move on to something else before coming back to it.
- *Allow time for incubating.* Start assignments early.

LEARNING BY OSMOSIS

CHAPTER 7

Quizzes, Tests, and Exams

When I'm working on a problem, I never think about beauty. I think only how to solve the problem. But when I have finished, if the solution is not beautiful, I know it is wrong.

—BUCKMINSTER FULLER (1895–1983)
American inventor of the geodesic dome

The only real differences between quizzes, tests, and exams are their percentage value of your entire grade and the frequency with which they are given. Quizzes are usually worth less than tests and may occur as often as once a week. A professor who prefers tests may have three or four in a term at "good stopping points" in the material. Exams, on the other hand, are the most comprehensive form of test and are usually given at the end of the course and sometimes during the middle of the term (midterms!). There will always be one professor who thinks the way to reduce class anxiety is to call the final exam "a quiz." However, the more frequently a professor tests you on the material, the more likely you are to learn it. In terms of the method for studying, all testing should be approached the same way. Of course, the best preparation for an exam is to stay on top of your course work during the whole quarter or semester, but this isn't always realistic or even possible. In the following sections, unless otherwise specified, tests will refer to any kind of testing.

STUDYING FOR TESTS

Top students often put less time into studying than the rest of us. Doesn't seem quite fair, does it? The difference isn't intellectual ability, it's their ability to organize efficiently and focus intently on the material they are studying. Anyone can do the same thing following these fabulous steps to better prepare for tests.

1. Organize.
2. Produce a game plan and study sheet.
3. Go over old tests.

4. Review class and text notes.
5. Practice problems.
6. Know your study sheet.

It looks like a lot of work, but it isn't really. The goal is to work smarter not harder. You can attack the whole list at once or take a couple of steps at a time. Not all the steps will be weighted equally in importance or time; these factors depend on the class and type of test you are facing.

Engineering students typically spend 100–170 hours (that's a week!) taking exams in earning their bachelor's degrees. (Higher end represents schools with quarter systems, lower end, semester systems.)

Step 1. Organize Your Troops (You and Your Notes)

You've gotten up early. You've primed yourself to hit the books . . . just bought a new highlighter, a waterproof felt-tip pen, and a decent stash of junk food to sneak into the library. You bike to the library, lock up, and secure a cubicle that looks knowledge inducing. Then you realize you've forgotten your exam outline—so to remember everything, take a quick look at the following checklist:

The Complete Don't-Leave-Home-Without-Everything Checklist

☐ Class notes and handouts
☐ Subject study guide (if available)
☐ Highlighter
☐ Pencils
☐ Big eraser
☐ Scrap paper
☐ Class textbook and class pack (even if you never used them)
☐ All problem sets and solutions
☐ Syllabus
☐ Test outline
☐ Calculator
☐ Extra lead or a pencil sharpener
☐ Sticky tabs (to mark important pages)
☐ Spare change for vending or photocopy machines

Wait! Before you head out the door (fully equipped) to study, if you need to spend some time getting organized, do it now while you are at home. Rearrange notes, homework sets, solutions, outlines, and any other loose items in an orderly form in your binder or folder. Now is also a good time to unearth notes you forgot you ever took (in a different notebook). Completing your notes and sorting the stack of information in a logical sequence before sitting down to study will save time and prevent a good deal of frustration.

The Top 10 Reasons Why Engineering Tests Are Better Than Liberal Arts Tests

1. No essay questions.
2. Your grade is determined by what you've learned rather than how well you can BS.
3. Grading is seldom subjective.
4. You get to draw pictures and use cool symbols.
5. Rarely will you be tested on readings.
6. A 65 percent could get curved up to a B+!
7. You are encouraged to use a pencil.

8. You don't usually have to outline engineering solutions before you start writing.
9. Some schools give more time for engineering exams.
10. You get to use your calculator.

Step 2. Be Strategic and Formulate a Game Plan

Tests are skirmishes! Exams are war! Studying for tests is a strategic exercise in preparation, especially when time is short (isn't it always?) and you have a sizable amount of material to cover. A good strategy considers three elements.

1. *How worthy is your opponent?* Size up the test. What are you in for? A rout or a walk in the park? In sizing up the test, you are anticipating how much time you should spend on and how deeply you should delve into each topic. Look back at old tests (your first test tells all) and talk to upperclassmen. Some professors will devote a whole class to reviewing for an exam while others will just remind you of the date. Always be on the alert for any hints the professor or TA might give regarding the exam.

A note about review or study sessions. Always try to go to the first review session a professor or TA offers. You may find the sessions very useful or a waste of time. It is best to attend a study session after you have done at least a preliminary review of the material so that you have a better idea of your own questions and the relevance of your classmates' questions. It may turn out to be more beneficial to go to the professor's office hours or make an appointment with the TA instead of wasting precious last minutes.

Besides learning as much as you can about the structure and nature of the test, how else can you derive information? Look at the table "Professor-Student Dictionary" on the next page, which provides some key words that should alert you.

2. *Anticipate the attacks.* What will be on the test? Look at how your professor grouped the classroom material and predict what types of questions will be asked. Consider the type of questions you would ask if you were teaching the class. Here are some clues to what will show up on the test or exam:

   - *What are the professor's favorite problems?* Think about any material or problems that really rocked his boat. Did your professor tell an anecdote or draw on the chalkboard—something he usually doesn't do? Sometimes professors will even warn the class to "look out for this one, you'll see it again."

   - *What kinds of problems integrate several concepts?* These problems are very popular with professors because they can test you on many things while asking only one question (or sometimes one question with eight parts).

☞ **Professors who make up their own homework problems** (instead of assigning text problems) tend to have similar problems on their tests.

- *What do the text and study guides deem important?* Look at applicable problems in Schaum's and the textbook. Common problems that might not make an appearance during class may occasionally show up on tests.

- *What do you not know well?* No kidding—it will probably be on the test. The material we often know the least about is the stuff taught right before the test.

Professor-Student Dictionary

Key Word(s)	What It Means to You
Open book	• Get familiar with the textbook's index! Put labeled tabs on pages with important graphs or tables. • Don't waste time during the exam flipping through the text! • This test won't be any easier than a closed-book test—if anything, harder. • Stick to studying concepts rather than details. Details are usually too much of a freebie. • When preparing for the test, spend less time memorizing and more time practicing problems. • Aim to not even open your book during the test. Often, you won't have the time to do so.
Closed book	• Use problem sets as an indication of which equations you should memorize. Professors will not usually expect you to have memorized random formulas on closed-book tests. • Try to find out if any charts, tables, and so forth will be provided. This will give you an indication of whether problems that require those will be on the test. • Work as many text examples as possible. It's a *closed*-book test for a reason!
Equations provided	• Guess which equations will be provided and what kinds of problems and solutions incorporate those equations. Study those especially! • When preparing for the test, spend less time memorizing and more time practicing problems.
Crib sheet (a.k.a. cheat sheet)	• Don't forget to allot time to write out your crib sheet neatly. • To get the most on your sheet, use a copier to neatly (and legibly!) reduce notes and good problems.

- Stick to studying concepts rather than details. Your crib sheet should cover the details.
- When preparing for the test, you can spend less time memorizing and more time practicing problems.

Proofs

- Be prepared to spend some time memorizing.
- Focus on fundamentals.
- Brush up on identities and mathematical relations that apply to the subject.
- Practice as many examples as possible. Often a professor has a limited number of proofs to choose from.

Cumulative

- Focus most of your studying on the *last third* of the course. The most recent material is usually stressed the most because earlier material has been covered on previous tests.
- When studying the earlier material, use old tests as a guide.

Take home

- Take home or take no prisoners means open books and notes (you can be tested on anything) and an open-ended amount of time (meaning it will take *much* longer than the regular test period to complete) within a specified period.
- If you have an ample amount of time to complete the take-home test, you may simply want to organize and lightly review the material before the test is given to you. Then delve into specific subjects once you have been given the test.
- Well before the exam is given out, check out any helpful library books relating to the topic.

Multiple choice or fill in the blank

- If you think some vocabulary will turn up, be prepared to spend some time memorizing.
- Look over boldface words in the textbook.
- Know equations for quick plug and chug solutions.
- Understand the units and dimensions you are dealing with to eliminate obviously wrong choices.

3. *Make a list of likely targets and deploy your forces accordingly.* Formulate your game plan by listing concepts, laws, and methodologies on a single piece of paper. Label each concept with the means by which to master it. Were there particularly good problem examples from the book? Explanatory handouts from the class or TA's section? Class notes that clarified an important topic? Problem set solutions that spelled out a procedure? Your strategy for better learning the material shouldn't be simply to reread the text chapters you've covered in

class. *The major part of the test will most likely cover specific concepts illustrated in problem sets.*

Deploy your forces accordingly by producing a sheet (or a limited number of sheets) that addresses your game plan and from which you will do most of your studying. Study sheets can include formulas, derivations, proofs, diagrams, definitions, steps to solve a problem, example problems, definitions, pertinent page references, and reminders to yourself.

Step 3. Review Old Campaigns (Tests)

Review both your own old tests and tests from previous semesters or quarters, if possible. Are there any additions you should make to your game plan and study sheet? Old tests are useful to you for two reasons.

1. *Old tests offer clues to the professor's testing style.* If you have old tests that are composed of analytical problems only, you shouldn't spend much time memorizing definitions or theory.
2. *Old tests are a great resource for extra practice questions.* Tests from past courses[1] may also include material you didn't realize was important.

> **The 80-20 Rule.** Twenty percent of the material in the course will account for 80 percent of the test problems and questions. This comes from an expansion on the Pareto Principle by Adam Robinson, cofounder of the *Princeton Review*. Pareto was a nineteenth-century Italian economist and sociologist who noted that from any group of objects, a small fraction contributes the most to the whole.

Step 4. Review Class and Text Notes

You should do most of your studying from your summary sheet, but reviewing notes will help you fit the theory and application puzzle pieces together for the big picture. Being able to understand and apply engineering theory will propel your knowledge from the short-term holding tank to long-term storage.

Class and text notes will reveal significant subjects or points that you might have missed earlier preparing your alpha[2] summary sheet. Add 'em on. You should now have a thorough estimation of the test material and how far along you are.

---

[1]See page 73 on the great student-professor debate about old tests.
[2]Before a product is released to the public, it often goes through many revisions. The first version of a large project (e.g., software) or part of a project (e.g., layout drawings for one part of a product) is often called the "alpha" version; likewise, the second version, after adjustments and corrections, is called "beta," and so on.

Step 5. Practice Problems

You don't prepare for a campaign by reading books; you practice what you think you will encounter out there on the battlefield. The bulk of your "study" time should be spent training. Work through drill problems. Look up references that explain the reasoning behind the method. Go through each section and make a list of all the problems you should practice. Pull problems from old tests, your study guide, or any homework sets for which you have solutions. Check off the problems as you work through them and make notes beside the ones that were particularly challenging or would make solid test questions. Work through the problems in a logical order and don't waste your time on problems without solutions!

Step 6. Know Your Study Sheet

When you cross off the final item on your game plan, it's time to learn your study sheet. Yes, this is boring, but after everything you've done, memorizing is a cinch! The best way to do this is to make yourself rewrite it from memory. It will take more than a few tries. An open-book test or an exam with equations included doesn't make you exempt from this step! A high recall speed during an exam will produce a sharper, more confident mind and allow for extra time at the end for checking over your work.

Should the All-Nighter Be an Option?
Turn to page 143 to find out!

> **Studying to Test Well versus Studying to Learn**
> Ideally, you want to be able to ace every test while accumulating omnipotent knowledge. In engineering, you can sometimes get away with getting pretty good grades when you don't really have a clue about what is going on. This is the studying shortcut version of the problem set shortcut (see pages 73–74) and can be accomplished by memorizing steps or methods instead of learning the cause-and-effect relationships. Absolutely try *not* to do this; if you skate through the principles and material now, it *will* come back to haunt you.
>
> On the flip side—if you are frustrated because you know every relationship and definition, and your tests don't reflect how well you really understand the material, then you need to concentrate more on general problem solving and less on the details.

CHARGE! THE TEST

The average student/faculty ratio at American universities is 15:2.

Keep in mind that the most important issue when taking any engineering test is time. So don't forget your watch! Some engineering professors take pride in testing your endurance and mental stability by delivering exams that require twice as much time and clarity as you are given. When heading into a test, remember:

The goal is to produce error-free work quickly and efficiently.

This takes practice, but not to worry—you will be given plenty! How do you start off?

When you get the exam:

 Take a quick inventory. Flip through it to identify the number of questions and estimate the amount of time you have for each. Also note the relative weight of the various problems.

Start with the easiest question and move to the hardest. The idea is to complete the questions that you know and leave extra time at the end for the problems that will take you longer to figure out. Consider it stretching for the run.

Try not to leave the highest weight problems until the end. This is the only time you probably should not leave the hardest problem to the end.

When tackling a problem:

 Read all *directions before jumping into solving the problem.* Engineering professors sometimes like to tack little short-answer questions onto the end of wordy problem statements.

Circle units. That way you won't forget to check if they are consistent.

- *Draw a diagram for clarity whenever you can.*

- *Plug numbers into equations at the very end of your solution.* Doing this saves time and writing, and reduces chances of human and decimal point errors.

- *Carry the units through the equations to avoid conversion errors.* This is extra writing, but the units will be self-checking.

- *Never leave a question blank!* Even if you have absolutely no idea what it may be about, jot down some equations, laws, or definitions that you think are in the ballpark. See pages 111–113 for help in reasoning through the tough ones.

- *Keep an eye on your equations and numbers while you are working through them.* If the equation doesn't look quite right, it probably isn't. Double-check as much as possible as you progress through the test in case there isn't time to go back and recheck answers.

It was the end of my sophomore year, I was taking dynamics, and I was really strug-gling. The concepts were tough for me, and homework really took a long time. I had managed to find a pretty good study group, but was still not feeling very confident in my skills heading into the first test. Other folks I was studying with were similarly nervous. The night before the big exam, we got together in the science library to study together, and began going through old homework assignments. Finally someone suggested that we do problems from the book that had not been assigned to practice. We chose a problem with a variety of weights and gears, and started work. About an hour and a half later, we were still wrestling with it. Finally, we all figured it out. With this boost of confidence, we headed off to get some sleep.

The next morning during the exam, looking through the problems, I saw a familiar diagram. The EXACT problem we had worked the night before was on the test! I remember looking around the test hall and seeing smiles on the faces of all my study group buddies. It was a weird bonding moment, knowing that our hard work the night before had com-pletely paid off.

(K.H.D., Biomedical Engineering '94, Brown University)

WHAT'S THE MODE OF ATTACK?—MULTIPLE CHOICE, SHORT ANSWER, OR LONG ANSWER?

Multiple Choice

Tests with multiple-choice problems can be good for the test taker or conversely detrimental. Check out the following test question:

A quartz piezo-electric crystal with a thickness of 1.5 mm and a voltage sensitivity of 0.055 $V \cdot m/N$ is subjected to 175 psi pressure. What is the voltage output?

(a) 99.5 V (b) 14.4 V (c) 0.144 V (d) 86.2 A

The good news: Multiple-choice questions give you the answer! Granted you may not know which of the four answers is the correct one, but if you are having a hard time getting started, they will provide useful clues. The bad news: Professors sometimes don't give partial credit on multiple-choice tests.

To be a master at multiple choice:

- *Look for clues in the choices before you work the problem.* The answers provide possible clues on the units and ballpark value of your final answer.

- *Eliminate answers to improve your chances.* Occasionally one choice is a freebie elimination. How can an output voltage be in amps? Cross out (d).

- *Check those units.* Incorrect choices are often traps for those who forget to convert.

- *Work the problem only as far as needed to select the answer.* If you are taking a "scantron" (a test where you fill out the bubble sheets), don't spend time documenting your process if you won't get credit for it.

- *Don't stress if your answer is slightly off.* The solution below gets 99.8 V as the final answer—what gives? Round-off error can throw you off slightly. Here the error occurred in the conversion from psi to N/m^2. Converting pressure to N/m^2, 175 psi = 1,205,750 N/m^2. This affects the final answer with $E = 99.47\ V$.

A quartz piezo-electric crystal with a thickness of 1.5 mm and a voltage sensitivity of 0.055 ($V{\cdot}m/N$) is subjected to 175 psi pressure. What is the voltage output?
→ *convert to metric!*

(a) 99.5 V   (b) 14.4 V   (c) 0.144 V   ~~(d) 86.2 A~~

$E = gtp$

$t = 1.5mm = 0.0015m$

$g = .055\ V{\cdot}m/N$

$p = 175\ psi = (175\ lb/in^2)\left(\dfrac{6.895\ E3\ N/m^2}{lb/in^2}\right)$

$\approx 1.21\ E6\ N/m^2$

$= \left(.055\ \dfrac{V{\cdot}m}{N}\right)(.0015m)(1.21\ E\ 6\ N/m^2)$

$= 99.8V \rightarrow$ (a)

Short Answer

Short-answer engineering tests can take one of two forms: qualitative or quantitative. Either should be easy if you've learned the theory and memorized the right equations. A qualitative problem from an Intro to C Programming class might be:

> What is the difference between &x and *x?

For quality answers, a one- or two-sentence answer with an example of each is the best way to go. Questions like this come directly out of assigned readings or class lectures. Your answer to the question above might read something like this:

> & is an "Address-of" pointer operator. Given a particular lvalue, it returns the memory address in which the lvalue is stored.
>
> * is a "value-pointed-to" pointer operator. This operator takes a value of any pointer type and returns the lvalue to which it points.

A quantitative chemistry problem might be:

> What is the pH of a solution if $[H3O^{+1}] = 7.83 - 10^{-6}$?

Short-answer problems that require calculation define "plug and chug"; there isn't much to figure out, just use an equation. This problem simply checks to see if you know the equation: $pH = -\log[H_3O^{+1}]$. Your answer should look something like this:

$$pH = -\log [H_3O^{+1}]$$
$$= -\log [7.83 \, E-6]$$
$$= -(-5.106)$$

$$\boxed{pH = 5.11}$$

Long Answer

Long-answer problems are the most notorious, the most frequent, and the most draining. Even when you do well, you can still feel like a stampede of elephants trampled over you.

> Physicsman is chasing Ann Tygravity when she appears to materialize on the roof of another building. He musters all his potential energy and takes a *horizontal* flying leap from the roof of his building to hers. Her building is on the other side of a busy street (20m) and 10m shorter. (a) What must be Physicsman's starting velocity to jump onto Ann Tygravity's rooftop? (Assume he needs 1m to land.) (b) How long will it take him to make the jump? (c) Will Ann Tygravity get away? (She needs 30s to get away.) (d) If Physicsman jumped at an angle 10° above the horizon with the same speed you calculated in (a), how far could he jump then? (e) Does he still land on the building (it is 10m wide)?

To make the long-answer problems less problematic:

🐾 *Draw a diagram!* A simplified free body diagram or a flowchart illustrating related items will help tremendously.

- Cross out parts of the problem that are unnecessary. Physicsman may be everyone's hero, but he won't help you solve this problem.

- Label the parts of your solution. It becomes more obvious if you answered all parts of the problem.

- Write down all equations you think apply. It helps having all the equations you need in front of you.

- Watch the clock. Budget your time so you have time to attempt each problem.

- Pull a stapled test apart. Flipping through pages is an annoying way to take a test (also annoying for others to listen to). If a problem takes up more than one page, keep the first page next to you as you start on the following page. When you complete a problem, retire it to a separate "out" pile.

- Cross out instead of erasing. You may want to refer back to it.

~~Physicsman is chasing Ann Tygravity when she appears to materialize on the roof of another building. He musters all his potential energy~~ and takes a _horizontal_ flying leap from the roof of his building to hers. Her building is on the other side of a busy street (20 m) and 10 m shorter. (a) What must be Physicsman's starting velocity to jump onto Ann Tygravity's rooftop? (Assume he needs 1 m to land.) (b) How long will it take him to make the jump? (c) Will Ann Tygravity get away? (She needs 30 s to get away). (d) If Physicsman jumped at an angle 10° above the horizon with the same speed you calculated in (a), how far would he horizontally jump then? (e) Would he land on the building if it was 10 m wide?

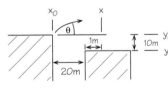

$$x - x_0 = v_{x0}t = (v_0 \cos \Theta_0)t$$

$$y - y_0 = v_{y0}t - \frac{1}{2}gt^2$$

$$= (v_0 \sin \Theta_0)t - \frac{1}{2}gt^2$$

(a) Horizontal jump: $\Theta = 0 \therefore v_{y_0} = 0$

so vertical distance $y - y_0 = -10m$

$$= (\cancel{v_0 \sin \Theta_0})t - \frac{1}{2}gt^2$$

$$-10m = -\frac{1}{2}gt^2 \quad \rightarrow \quad t = \sqrt{\frac{(20m)}{g}}$$

horizontal distance jumped: $x - x_0 = (\cancel{v_0 \cos \Theta_0})t$

$$(20 + 1)m = v_0\sqrt{\frac{(20m)}{g}} \quad \rightarrow \quad v_0 = (21m)\sqrt{\frac{g}{(20m)}}$$

$$\boxed{v_0 = 14.7 \text{ m/s}}$$

(b) From (a): $t = \sqrt{\dfrac{20m}{g}}$ $\rightarrow$ $\boxed{t = 1.43s}$

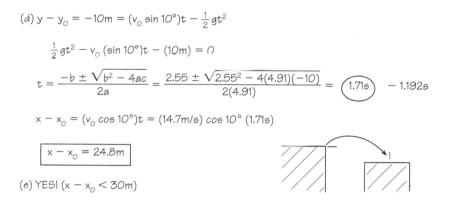

$\theta = 10°$

20m   10m

(c) No. He's got lots of time $\approx 28.5s$

(d) $y - y_0 = -10m = (v_0 \sin 10°)t - \dfrac{1}{2}gt^2$

$\dfrac{1}{2}gt^2 - v_0(\sin 10°)t - (10m) = 0$

$t = \dfrac{-b \pm \sqrt{b^2 - 4ac}}{2a} = \dfrac{2.55 \pm \sqrt{2.55^2 - 4(4.91)(-10)}}{2(4.91)} = \boxed{1.71s} \; -1.192s$

$x - x_0 = (v_0 \cos 10°)t = (14.7m/s) \cos 10° (1.71s)$

$\boxed{x - x_0 = 24.8m}$

(e) YES! $(x - x_0 < 30m)$

Four Things to Consider If Your Answer Seems Way Off

1. **Check your units.**

    Did you forget to convert? Did you remember to account for prefixes like giga or milli? Does the answer seem wrong when it isn't because the test question uses English units when the class examples and homework problems with which you are familiar use SI units?

2. **Recheck any numbers you took from a chart, table, or graph.**

    It is *very* easy to misread values on cramped photocopied charts and tables. A minute interpretation difference of a log-log or semi-log chart can produce a monumental difference in the final result.

3. **Could the answer actually be correct?**

    Maybe it would look more realistic if you converted mols back to kmols (1000 mols = 1 kmols). Or perhaps your professor is using an extreme situation for a "more interesting" problem.

4. **Write a note to maximize partial credit.**

    If you are out of time and the answer violates your engineering intuition, make a note along these lines: "The answer appears to be off, but I don't have time to find it. A more reasonable answer might be. . . . "

Haven't a Clue . . .

You've hit a question and you have *no idea* where your professor dug it up. The hardest questions on the exam are often not as awful as they appear—if you can decipher the question and make the connection to the material. Tough problems are frequently some application of important theory that was not covered specifically in a class example. Something somewhere applies!

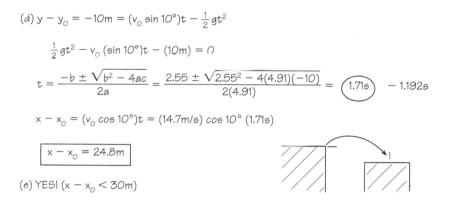

OK, the most important thing to remember when you have no clue:

Never EVER leave a question blank!

If you leave a question blank, you are telling the professor that you don't know anything . . . which is always untrue. So write down an equation, a definition, or some sort of reasoning that shows that you do know something about the concept. For a 15-point problem, 2 or 3 points is always better than zero points.

Just because you have no earthly idea how to do a problem on the first pass doesn't mean the solution won't come to you while you are working on other problems. Sometimes all you need is a warm-up problem to get you thinking or another question that might trigger the remembrance of an equation or method that had previously eluded you.

OK, now you have completed the whole test except for the hardest problem. Now what? Here are few tips that will help you put up a fight:

1. *What are the variables involved?* Rewrite them and get rid of some of that blank space on your test paper. It is easier to take an inventory when unknowns, constants, and knowns are listed. What equations and/or methods that you covered employ these numbers or variables?

2. *Look for key words in the question.* Wording or terminology can often point you in the right direction. Even if you don't hit the correct answer, you'll probably get points for being in the right ballpark.

3. *What has the professor not covered with the other questions?* Use the process of elimination. Reason through the test and think about what else is likely to be on the test, but isn't.

4. *Is there a key chart, graph, or table that may fill a void in the information?* Double-check whether any charts are included with the exam or if the professor has written any notes on the board.

5. *Just start writing.* Actively tackling the problem can trigger your memory of a method or a similar example. Should you run out of time, you'll at least have work down on the paper.

6. *Are there any common elements with a similar or related problem?* Does this problem sound sort of like another problem you studied? Writing down a general formula might jog your memory to other uses.

7. *Examine the units for clues.* What quantities are you working with and how might they be related? For example:

   MPa (1 MPa = 1 megapascal = 10^6 pascals = $10^6 N/m^2$ = 1 N/mm^2)

   could be stress or pressure and they are both equal to force divided by area, which is equal to mass times gravity divided by length squared.

8. *Rewrite the problem in your own words.* You may learn things you didn't pick up when reading through the question. This will also ensure that you understand exactly what is being asked.

9. *If you really get stumped at one part, use poetic license to fill in the blanks.* State in your own words that you are not sure how to complete the problem given, but with a sign change or given a particular variable, you are able to work the problem and produce the following result.

Of course, your solution is incorrect, but most of your method will be correct. You probably just forgot an integral or simple equation and will only lose minor points.

10. *Could you maybe derive an equation you've forgotten?* In part (c) of the sample physics problem on page 111, you could have more efficiently used $x - x_0 = (v_0^2/g)\sin 2\theta$ instead of working through the equations of parts (a) and (b). Our test taker, in fact, derived it by working through the solution the "long way."

> ☞ **Worse comes to worst.** If you have a test or exam that is just absolutely and utterly horrendous—due to a miscalculation on the professor's part, certainly not because you didn't study—take comfort in that *it is a horrendous exam for everyone* and it will soon be over. Imagine this test as a take-home exam and you had three days of it instead of an hour or two.

GETTING THE TEST BACK

Whether you were victorious or suffered a humiliating defeat, don't forget to look over the professor's comments and place the test somewhere retrievable (i.e., not the recycled paper bin) for final exam time.

> ☞ **If you think there may be a grading error,** take your test home and work the problem again—and then make an appointment with your professor. If you used a different (but correct) method to solve a problem, take the reference notes or textbook with you to your meeting. Professors are bombarded with students claiming they have been unjustly wronged; your tact and conscientiousness in handling the situation will likely gain you their respect.

Hard to imagine in the middle of exams, but engineers like school! Statistics show that after you start working, there is an 82.5 percent chance that you will take at least one more class (while you are working!).

CHAPTER 8

A Crash Course in Engineering

To engineer is human.

—HENRY PETROSKI (1942–)
American civil engineer and author

As if your classes weren't already a crash course in engineering! Okay, but let's change the perspective. If you take Turkish next semester you will probably do most (if not all) of the following: go to class (more often than not), do homework assignments, and listen to audio clips at language lab once a week. Determined students will try their best to practice new vocabulary in class, and listen to Turkish radio livestream while working on other homework (like a statics problem set!). But how much better could you learn the language if you went and hung out in Turkey for a quarter? Probably pretty well—and on top of that, you are able to see the point of all those trips over to the language lab.

Engineering isn't so different—except the immersion experience isn't so far away. Right now you: go to class (more times than not), do your problem sets, head over to the bio/physics/chem/materials lab once a week or so, try your best to follow what's going on in class, and maybe even ponder zero-force members during your Turkish professor slideshow on architecture of the Ottoman empire. The good news is that you won't require a passport to put your engineering skills to work as an intern or co-op, or on a research project.

Where engineering *does* differ from a language is with prerequisites. If you want to learn Turkish you can simply start going to Turkish class. If you want to learn about structures, for example, it is likely you will spend one or two years wading through prerequisites before you take your first class in structures! Because of this, it isn't always easy to stay motivated—especially if you can't see the connections of what you are learning to real life. Internships, co-ops, and research opportunities allow you to jump into engineering at any stage and see the value of the prereqs and classes, while applying some of your own technical know-how.

A Crash Course
in Engineering

The grade point averages of co-op students at universities with formal co-op programs are significantly higher than those of students who do not participate in the co-op programs.

What is the difference between co-ops and internships anyway? Both internships and co-ops are opportunities for "real world" experience—which, of course, has its benefits and drawbacks. But they have significant differences that aren't so obvious. Internships are *usually* summer jobs working in a professional area that may or may not be related to your major or field of study. Co-ops are similar to internships in that while you may start with little (or no) on-the-job experience, you will end with lots—experience that is proven to be very valuable when you look for a full-time job after graduation.

The similarities generally end there. Co-ops are typically taken during the regular school year for a semester or quarter—or longer. Co-op is short for "co-operative learning/education," and as the name reflects, co-ops usually have an educational bent to them. Whereas an internship may or may not be in a field related to your major, co-ops almost always are—many universities will give academic credit for participation. Some schools even have a classroom or seminar component where you meet weekly with other co-op students and submit progress reports and evaluations. Co-ops tend to pay a bit less than internships, and participation in them during the school year may result in the extension of your studies.

That said, why might you choose a co-op over an internship? Two reasons. Companies that offer co-op opportunities rather than internships often recognize that your experience with them is an important component of your education, and may make a stronger effort to give you a "good project" to work on. Also, there tend to be fewer co-op students at any one time than there are summer interns, which means you may get more guidance and advice from superiors.

Hmmm . . . still unsure why you would want to give up late nights and late mornings? Consider what an internship or co-op offers:

- *A job now!* That's right—a vacation from tests and exams for a whole quarter and instead of paying, you get paid! Okay well, an internship or co-op certainly isn't a vacation, but it will be a change of pace and a chance to test run your future plans and ambitions.

- *A job at graduation!* National statistics show that 63 percent of co-op students receive an offer for full-time employment from their employer at graduation. Not only that, while it takes the average engineering grad 6 to 15 months to find a job, the average time for co-op students is 3 months!

- *"Now I understand!"* So vectors didn't mean much in 2-D (on the chalkboard), but in 3-D you can't imagine life without them.

- *A glimpse into the future.* Internships and co-ops take away some of the mystery (and anxiety) about what the heck you are going to do with your life. You get a chance to figure out things like what sort of field you want to work in ("But it was hard enough to choose EE, and now you're telling me that there are hundreds of subspecialties?"), your ideal work culture, location, and benefits.

It's only a test run. Internships and co-ops are jobs *with* all the worries of a job, but after a few months it will end. After graduation you don't have the same opportunity to try jobs out before dedicating more than half of your awake hours.

Assuming You Are Now Interested in a Co-op or Internship—
What Is a Likely Next Step?

1. Finding an internship or co-op.

In some ways an internship or co-op is the Mini Me version of a job. You need to do most of the things associated with job hunting, but on a smaller scale: pull together a resume that reflects your brains and talent, figure out where you want work, interview, and (if appropriate) negotiate your terms. The best place to start is with your university: check out the career center to see if your school has formal internships or co-op programs. Even if it doesn't, it still should be able to offer you some advice. Other resources: department announcements, professional society webpages, and other students who have been interns or have co-op experience.

What should you do if your school doesn't have a co-op program? Many schools allow students to take a leave of absence or "stop out" for a variety of reasons—not all of them arising from family or personal difficulties. Co-ops and internships are available year-round at almost every major technology company. To best learn about university regulations regarding a do-it-yourself co-op and how to navigate the accompanying paperwork, try any or all of these four resources: your university career center, the registrar's office, the Web, and your advisor.

---

2. Choosing an internship or co-op.

There are four L's to consider when you are comparing your offers:

 *Gotta **like** it.* Which company do you like the most? No kidding. How well did previous interns rate their experiences?

 *Logic says to think about **logistics**.* Do the pay, location, and hours work for you? According to Penn State's Cooperative Education Office, students participating in co-ops earn 50 to 75 percent of their full-time entry-level engineering peers. Some employers also offer tuition assistance, stock options, and/or benefits.

 *Thinking about **later**.* Think this might be a place you want to work at later?

 Yawn. Yes—LLLLLLearning! The money you make will be good, but you could probably make more if you dropped out of school and worked somewhere else altogether. You are really there to learn about getting the hang of it all. So you need to determine: Will you be left to sink or swim? Will you be assigned a mentor? Does it look like a place where people will stop what they are doing to explain things to you? Have they had interns before? Does the project you will be assigned to be part of sound interesting?

> **Thinking about a school-year internship or co-op?** Beware of the hidden driveway called your registration status. Check out what your job means to your financial aid status, dormitory contract, and library/computer/sports center privileges?

3. Reporting for work.

So now you've got a job—how to get the MOST out of three months?

 Don't worry! You aren't expected to know everything. That's why you are still in school.

 Social interaction is key. Everyone knows this already (even you), but no one ever says it directly. Get to know everyone from the janitor to the CEO. Network and make the time to get to know all the groups and individuals you can.

Superimpose experiences. While internships and co-ops sell themselves on the technical stuff you do and learn, there is also great value in learning about your co-workers' experiences. Some of the people you will be working with were in your shoes not so long ago and can give incredible insight into all sorts of issues you might not think to even ask about—company culture, balancing family and careers, regrets and smart moves, etc. Why did the engineer you report to leave his last job to come here? What classes did your project manager wish that she had taken when she was in university?

Curiosity won't kill this cat. You *do* want to exhibit your knowledge and ability to reason things out on your own, but that may be tough to do without a few explanations here and there. Ask questions. Be proactive in getting answers. Don't be afraid to ask for clarifications and examples.

Quantify your experience. At the end of a course you usually have an exam or a project where you are supposed to apply everything you have learned, right? Internships and co-ops don't always have that opportunity (that's a new way to think about exams, isn't it?). Keep track of the things you are learning. From how *not* to put together a circuit board (oops . . . but it was only an evacuation—no fire trucks!) to using material guides for snap fits (you mean we don't have to derive equations—it's *right there?!?* COOL.)

How to say I am an engineering student *in Turkish:* Mühendislik öğrencisiyim *(closest anglophone pronunciation: mu-hen-dis-lick uh-ren-gee-cee-yim)*

RESEARCH ASSISTANTSHIPS
(RAship s)

Basic research is what I'm doing when I don't know what I am doing.
—WERNHER VON BRAUN (1912–1977)
German-American rocket engineer

Nine to five isn't really your style. You like lab classes, but wish the findings weren't always so predictable. Field trips are still one of your favorite school events. Or maybe you just simply *like* research.

Research assistantships are exactly that—you assist with a research project usually (but not always) at your university. You will typically work with a professor, post-docs,[1] or graduate students sometimes for school credit ("independent study" or "directed study"), sometimes for money or tuition—and sometimes neither. As an undergrad you will likely be assigned the grunt work: cleaning tanks, data entry, tagging samples; and while this may not always be fun or glamorous, it is an important aspect of every project. If you are thinking about graduate school or a research job after you graduate, research experience is invaluable. What makes the grunt work valuable?

- *You will likely be in over your head.* This is a benefit? YEEsh! If you decide to do more research in the future, you will experience this feeling many times. And not just in research, some engineers deliberately choose jobs at companies where they have the ability to work on completely new projects when their old ones end. Participation in any project (research or industry) always starts with feeling that there is no way you will ever learn everything you need to say something at a group meeting. You will surprise yourself how quickly you will learn and contribute.

- *You are in the know.* While you might not single-handedly come up with a new model for turbulence (you might though too!), evidence suggests that you will understand the culture of research—like how to set up experiments and why attention to detail (like dating all of your notes) pays off. Not only that, you will learn vocabulary important to the field and how to run cool machines and equipment.

- *Insight into graduate student life.* Most students who do engineering research as undergrads are individuals thinking about going to graduate school. In the same way internships or co-ops are glimpses as to what an engineering job is like, research experience is the crystal ball for graduate school. You don't need research experience to be accepted to a graduate program, but before you devote another year (or *many* more) to more education, it's a good idea to take a gander as to what you will likely be up to.

- *Your professor was the national tiddly-wink champ eight years ago?* Students with research experience often say that one of the most valuable aspects of it was getting to interact with graduate students, faculty, and staff in a less formal setting.

At the beginning of my sophomore year, I decided that I wanted to do some research. I was clueless about the different areas within my field, so I just picked the most approachable professor I had taken a class from, and stopped by his office one day. Despite his friendliness, I was so nervous talking to him that I think I was stuttering. Despite the painfulness of that first approach, the research I started then was incredibly important to my development as an engineer and later as an academic. Now that I am a professor, I try to be as friendly as possible to students that approach me about doing research. But they never seem to be as nervous as I was back then!

(A.O., Mechanical Engineering '94, University of California–Berkeley)

[1]*Post-docs* is the shortened form of "post-doctoral fellows/scholars"; folks who usually love research and can be considered something like a cross between a PhD student and professor/lecturer.

Decisions made easy. Even if you are undecided about a major, research assistantships can help you figure out what you like and don't like. For instance, a research position where you are measuring the height of cypress tree knees in a swamp three days a week may teach you that even though environmental engineering isn't for you, you love field work.

 Turn work-study into *research*-study. Many universities allow students to assist professors as part of their work-study program. Talk to your work-study or financial aid office about this possibility.

---

1. **You're sold. Finding an RAship:**

    You particularly dug a class. If there is a class that you really liked, go talk to the professor about research opportunities in his or her lab—or with another appropriate group.

    Ask your TAs about their research. If your TA is a graduate student, he or she is probably working on a research project. Ask if a lab website exists outlining his or her work, and if you can come look around his or her lab.

    Surf the Web. Almost all the research being done at universities is covered on the Web. If you find a topic that interests you, don't be shy about approaching professors you don't know. Even if there isn't a position for you, they will be impressed with your motivation and give you some good suggestions.

Mmmm...Nothin' like a good cup'a *Turbulence* in the mornin'...

Look for an undergraduate research office at your school. This office should have information on existing programs such as Research Experience for Undergraduates (REU) sponsored by the National Science Foundation. Some universities have funds available for students to do research on a topic of their own choosing with guidance by a faculty member.

An REU for You?

Not so long ago, the National Science Foundation started a program called "Research Experience for Undergraduates" (REU) where students are able to do research in all fields of science, math, and engineering typically for 10 weeks during the summer. There are more than 100 sites (over 50 in engineering) at universities across the country. Each site takes approximately 10 students who are assigned work on various research projects (depending on the university's current projects and your interests) with professors, graduate students, and post-docs. In all cases, students receive stipends and, in some cases, assistance with housing and travel expenses. If you are interested in research, REUs are an excellent place to start as the programs are designed specifically for undergraduates; therefore, you don't have to have a ton of technical knowledge already. Universities with REUs are encouraged to provide seminars, lunch meetings, social functions, and mentoring—things you may not be able to find in a traditional research setting! More info on sites and projects can be found at:

http://www.nsf.gov/home/crssprgm/reu/start.htm

 Don't rule out internships or co-ops. Internships and co-ops are also available at corporate and national laboratories.

 Research doesn't happen just at universities. Major (and minor) corporations, government laboratories, and non-profit organizations do research all over the country.

 Treat it like a job hunt. Be prepared! Have ready your resume and a smart answer to: "Why are you interested in this research?"

2. Everyone wants you to work with them! How do you choose the best RAship?

 Check out the atmosphere! Is the laboratory quiet with reclusive graduate students who hardly know each other? Or do you see evidence of collaboration and interaction?

 Talk to past RAs. Past RAs can tell you what to expect in terms of quality of work, hours, professor moodiness, etc.

 Talk to the faculty member and/or graduate student overseeing the project. How interested are they in your personal education? Will they take time out to explain things to you?

- ● *Go to a group meeting.* Many professors have weekly or biweekly group meetings with their graduate students where they present their recent findings and present problems for suggestions from the group. Group meetings are great opportunities to check out group dynamics and level of professor commitment, meet people, and learn about the different projects going on.

- ● *How well formed is the project that you would be working on?* Some people like to jump into research problems that are clearly defined with the next steps naturally following. Others prefer the challenge of reasoning through experimental design and troubleshooting along the way. Many students who jump into research projects for only a quarter or semester become frustrated by the lack of definition and support.

3. **Now you are a researcher. How do you get the MOST of your RAship?**

- ● *Stay with it for more than a quarter or semester if you can.* It can be draining on your available time outside class, but faculty members and grad students will probably invest more in bringing you up to speed if they think you will be around longer.

- ● *Write a paper or present a poster.* Research findings are disseminated through conference posters, talks, or proceedings, articles in professional journals, and laboratory technical reports. While this may be more work than you want to do at the end of your RAship, it is *very* good publicity for you and nice closure to the experience.

- ● *Chat up the gatekeepers.* You will learn most through informal conversations. How does this *whole* experiment work? How do you know how often to collect data? Ummm . . . What exactly is digital particle tracking velocimetry? And what do you do with all that data? You will learn which decisions have theoretical basis and which do not. On top of that, grunt work is frustrating if you aren't able to unlock the key to why it is so necessary to the project.

- ● *Read.* Ask if there is any background material that will help you understand everything better.

- ● *Keep a research journal.* Keep track of acronyms. List your experimental setup (so it can be reproduced). Write down questions to ask later.

A preliminary study of undergraduate engineering students who participated in research showed that they were more likely to stay in engineering, more likely to pursue a job in engineering after graduation, and more likely to go to graduate school in engineering than national averages.

CHAPTER 9

When the Going Gets Tough . . . Dealing with Ruts and Unmarked Pitfalls

I felt a great disturbance in the Force.
—Sir Alec Guinness's character Obi-Wan
in the movie *Star Wars* (1977)

Every college student in every major passes through an academic or emotional rut at some point. Engineers seem to fall into more than their share, not because we are uncoordinated (hmmm), but because we *care more*. Engineering students (and maybe a few premeds) tend to be more concerned about their grades, what they are learning, how much they are remembering, and how all the theories fit into the Big Picture. Add to that, they already know that engineering isn't easy and the more retained the better off they'll be.

When floundering in the middle of a rough week, it can be perplexing to imagine that college is supposed to be fun. At the end of the quarter or semester when you look back at the work you have managed to complete, you see that it *is* worth it. And that doesn't even compare to the glee you feel when picking up your diploma after years of endless problem sets, programs, and projects.

There are three central things to keep in mind when the ground takes on a negative slope:

1. *Balance.* Have other active interests besides school (TV doesn't count).
2. *Take care of yourself.* Try to get a reasonable amount of sleep.
3. *Keep at it.* Playing Graham Nash's "Chippin' Away" in the background helps.

You can pretty much count on the fact that if you are having a tough time coping with the workload, grades, knowledge assimilation, and life, your

126

When the Going
Gets Tough . . .
Dealing with Ruts
and Unmarked
Pitfalls

classmates are too. Some pitfalls or ruts can be minimized or avoided altogether; others unfortunately cannot.

Next is a quick overview of the biggies and some possible solutions.

RUTS

Burnout

Burnout comes in many forms but is most often characterized by a lack of motivation. Additional symptoms can include academic disinterest, exhaustion, sleepiness, and procrastination like never before. Burnout can result from overdoing it the previous quarter, taking a required class you aren't interested in, a bad professor, too many nonacademic commitments and demands, or frustrating work arrangements (teams for math homework?!).

Burnout Solutions

- *Choose classes wisely.* If the semester looks like it might be intense, try to take a fun elective. If you can't take a free elective, try to pick an engineering elective that offers something different from the other classes. For example, if it looks like all your classes will be lecture and problem sets, try to take a lab.

- *After your freshman year, try to take hard or work-intensive classes first semester or over the summer.* The first semester or quarter after summer you usually return to school ready to hit the books; the second semester or second and third quarters you return to school wondering where the vacation time went. Summer school may be taught at an accelerated rate, but the atmosphere is often more laid back because everyone takes fewer units.

- *Reward yourself for a job well done.* Take an afternoon off or peruse and download some new music.

- *Find a compatible person with whom you study well.* Your study mate doesn't even have to be an engineer.

> **Study groups aren't always successful.** The "more the merrier" and more merriment can mean less work is accomplished.

- *Maintain other interests.*

- *Take study breaks* (see page 129).

- *Change your regular study routine.* Try studying in a new place, perhaps in the art library or a different study room. Buy some earplugs and head for a coffee shop or try to listen to music while studying.

To write a compact disc (CD), sound must be converted to digital code by sampling the sound waves at 44.1 kHz (thousand samples per second) and converting each sample into a 16 bit number. It requires almost a million and a half bits of storage for each second of stereo hi-fi sound.

Sleeping All the Time

127

When the Going
Gets Tough . . .
Dealing with Ruts
and Unmarked
Pitfalls

Caffeine just doesn't work anymore? Napping may not be so bad if it allows you to work at odd hours when you are most productive. Problems arise if you need more than one nap a day or if you are always tired and unproductive.

Solutions to Your Need for Naps

 Break the cycle. Go cold turkey and go to bed at a decent hour.

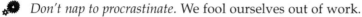 *Change your eating habits.*

Set regular hours for yourself and stick to them. Tell friends the latest time they can call you (then turn off your phone because they will still call). You will be more productive and time-efficient with your schoolwork when you have a bedtime.

Don't nap to procrastinate. We fool ourselves out of work.

Exercise more.

Keep an eye on yourself! Have too many late nights caught up with you and this is something that will pass with a more regular sleeping schedule? Or have you been feeling like this for a while? Persistent tiredness can be a symptom of depression or a physical ailment.

Too Much to Do

Napping? Who has time for napping? Even if you could nap—which you can't because you can't sleep, because you're not sure if napping is a luxury you should allow yourself, and anyway you would dream of engine cycles playing like film loops—but you *can't* because there are still a million things to do. Whew!

Solutions to Help Slow Down

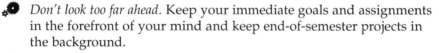 *Don't look too far ahead.* Keep your immediate goals and assignments in the forefront of your mind and keep end-of-semester projects in the background.

Remember that anxiety is a wasted emotion.

Cut down on your caffeine consumption. People who consume too much caffeine tend to be running a million miles an hour without focus or completion of anything. Read more about caffeine and kicking the habit starting on page 136.

 Are you multifunctional? Organize notes and punch holes in handouts while you are on the phone, take a book to the gym to read on the stationary bike, or review solution sheets while waiting for an appointment.

 Reevaluate how you spend your time. If you think there isn't any more you could possibly do, accept that you are doing the best you can and be proud of yourself.

128

When the Going
Gets Tough . . .
Dealing with Ruts
and Unmarked
Pitfalls

 Think you are just inefficient? Try keeping a time log to see how you are spending your time. Also peruse "Advice from the experts: How to make time fill the work allotted" on page 96.

Keep things in perspective. Bad grades stink, but things could be much worse.

Talk to peers who seem well adjusted and happy. How do they do it?

Keep your batteries charged and motor running. Chapter 10 looks at the physical and psychological symptoms of stress and anxiety and how to deal with it (see page 141).

TV and the Wednesday Night Beer

Television is bubble-gum for the mind.
—FRANK LLOYD WRIGHT (1867–1959)
American architect

This pitfall also could be called "Getting Sucked into Things You Really Don't Have Time For." Procrastination is one of those things that when it rains, it pours; it is difficult to procrastinate only a little bit. One beer turns into three or four, and a half hour of prime-time TV turns into 3 hours. Procrastination can make an assignment that should require 2 hours to complete into one that takes the whole night.

Solutions to Beat Procrastination

Tape shows. Tape just the television shows you want and watch them later at a study break or when you have time. That way you can fast-forward through commercials, too!

129

When the Going
Gets Tough . . .
Dealing with Ruts
and Unmarked
Pitfalls

 Set and meet targets. There is always some work you could be doing with engineering, so there never really is a time when you can get your "work done early and go out." Give yourself a target goal (like finishing all the problem sets due for the next two days and start on the one due Friday), so you can occasionally go out during the week and watch the full prime-time marathon if you want. Slack-off nights, when you can afford them, will keep you from getting burned out.

Go to the library, coffee house, study lounge, or picnic table on the lawn. Social studying is better than not even looking at a book. Getting out of the house gets you away from phone calls, bad-influence roommates, and other distractions.

Set blackout hours for time sinks like e-mail, music downloading, and solitaire. Regulate time sinks to specific hours—see if you can do e-mail for only 15 minutes in the morning and 15 minutes in the late afternoon. Besides, quick responses to e-mail and too many hours Napstering or playing with electronic cards doesn't reflect well on your social life.

Have productive study breaks. The key to a good study break is feeling sufficiently refreshed to tackle the books again. Try to set a time limit for your study break and then reevaluate how you feel. It's OK to extend your break *if necessary,* but it shouldn't be more than double the time you originally allotted.

The Best Study Breaks Are Those That Offer:
- *A change of environment* like a trip to the athletic center or cooking a real dinner (not from a box). Walking from your desk to the soda machine in the basement usually won't do it.
- *A jog down memory lane* such as writing a long e-mail or calling a friend from home.
- *Increased adrenaline* like playing intramurals or finding a date for a weekend party.
- *A glimpse of life off-campus* like watching the news, volunteering as an after-school tutor or mentor, or going to the store.
- *A change of mental scenery* like flipping through a magazine or reading a chapter from a good book.
- *Stress relief* like playing a musical instrument, hanging out with friends, or taking a dog for a long walk.

PITFALLS

Computer Hell

Aaghhhhhhhhhh! The server went down. Disk quota exceeded. ERROR. Offending command: bin obj seq. Type = 128. OFFENDING! Just *who* is offending WHOM? System crash and you haven't been saving every two minutes. It's

130

When the Going
Gets Tough . . .
Dealing with Ruts
and Unmarked
Pitfalls

not supposed to do this!! There really isn't anything you can say to comfort someone who just lost a valuable file or program. What makes computer disasters even more frustrating is that they usually can be avoided.

How Not to Let Your System Crash

 Start all programming assignments early. There are always bugs and better ways.

 Save often. Be paranoid.

 Revisit computer lab hours. Plan accordingly.

 Be one with your system. Understanding your university's operating systems (such as UNIX) makes life remarkably easier. Look for info sheets and check for "Getting Started with . . . " seminars.

 Invest in file recovery software. It doesn't always work, but sometimes it does.

 Back up. Back up often. Have backups of all programs, operating system disks, and installation disks for your own computer. Back up your files at least once a week—don't forget to back up e-mails, bookmarks and other program data.

 Always have one extra print cartridge and pack of paper. You always run out 10 minutes before the deadline.

 Know someone whose system is compatible with yours. This is important in case your computer isn't functioning properly and you need to get an assignment in or send an important e-mail.

131

When the Going
Gets Tough . . .
Dealing with Ruts
and Unmarked
Pitfalls

 Buy lots of disks and a portable hard-case disk carrier. You can never have too many disks. Leave several disks in the box in your backpack.

⚠ **Protect your disks!** If you don't have a transporter box, use a ziplock or an envelope. Funky things happen in backpacks.

 Label your disks with contents and your contact info so you know what is on the disk and someone else knows whom to return it to. You will likely forget more than one disk in a lab computer.

 Invest in surge protectors. Double-check that they are computer grade and don't forget to protect peripherals too!

---

*I*t was 3 A.M., and our exhausted group was on the last problem. We could think of no way to solve the quasi one-dimensional compressible nozzle flow problem of 10 equations and 10 unknowns other than using Mathematica. After about 5 hours of hacking a Mathematica script, we were ready. It had to work. Ten equations, 10 unknowns, what could go wrong? Nothing, except for the all-telling Mathematica error: "The equations you are attempting to solve appear to involve transcendental functions in an inherently non-algebraic way." Since by this time it was 8 A.M., we had to give up and go to class at 9 A.M.

After class, the three of us stuck around to ask the professor about this problem. He was a cheerful young professor, always willing to help and very approachable.

One of us proclaimed, "We were up all night and we couldn't solve this problem! Ten equations and 10 unknowns? We tried Mathematica on the SGI's and we still couldn't solve it."

Our professor just looked at us, paused for a moment (as if he'd been waiting a long time to say this), and pointed to the HP calculator we were using. "How much memory does your calculator have?"

Confused, one of us replied, "64 kilobytes. . . . "

Then, the professor raised his arms, and rolling his eyes about jokingly, exclaimed, "Why couldn't you solve the problem on that thing? You could put a man on the moon with your calculator!" And he was right. The Apollo 11 lunar mission used the most advanced computers of their time, and they were nothing compared to the HP calculators we use today. Somehow, his comment made us rethink the problem and not jump straight to the computer to attempt its solution. With that, we were able to solve it on our own, WITHOUT a computer.

O.B.F., Mechanical and Aerospace Engineering '95, Princeton University)

Can't Get into a Class You Need to Get Into?

This can be particularly frustrating. If you have a legitimate reason why you should be in the class, go talk to the professor. It might be a wild-goose chase, but if you are polite and *persistent* you will always get in. For some oversubscribed classes, professors will ask hopeful students to fill out a registration form with a question like "What do you hope to gain by taking ChemE 103?" Write a lot! Go all out! This little essay may determine whether or not you get into the class.

132

When the Going
Gets Tough . . .
Dealing with Ruts
and Unmarked
Pitfalls

End-of-the-Term Slam

Dead Week (a.k.a. Study Week or Vac Week) is a week at the end of the semester or quarter when professors are not allowed to teach new material or assign additional homework because students should be devoting their time and energy to preparing for finals. No one has yet figured out why many schools don't have a Dead Week for engineers. Who is dying if not the engineers?

Whoa! The workload is *increasing*. You're not imagining it. The professor probably realized she was a bit behind in the class schedule and material. You'll hopefully have read the assigned readings and have kept up so that your lack of time and abundance of stress won't be quadrupled in an attempt to have a strong end of the term.

Surviving the Slam Solutions

- *Keep a running list of assignments to complete and things to do.* You should give emphasis to priority items and how much time you can spend on each.

- *Put extracurriculars on hold.* Skipping morning runs for one week is probably good for you. Don't feel bad about returning low-priority phone calls when exams are over.

- *Think ahead.* Buy study foods, exam booklets, highlighters, paper, a nice pen, and so on ahead of time. Do your laundry so you'll have clean clothes. It all helps to get yourself into a work mode.

- *Eat easy-to-fix meals.* If you can't afford to grab quick bites at the school cafeteria, keep lasagna in the freezer and sandwich stuff in the fridge.

Adverse Advisors

Advisors are most often moonlighting professors. When you consider everything an average professor is expected to do, it's not so hard to understand why advising undergrads may fall pretty low on their list of priorities. Often, the difference between a good advisor and a bad advisor is timing. But, if it seems like your timing is consistently bad, here is some advising advice:

- *Be your own advisor.* Read the course manual and your curriculum sheet thoroughly. If you have questions, talk to a well-liked professor in your department.

- *Talk to upperclassmen.* Find out about good classes to fulfill requirements. Ask for recommendations for good upper-level classes that match your interests.

- *Call the registrar's office for university policy questions.* How are transfer credits handled? Are students in your major allowed to take classes as pass/fail? Often the university registrar knows the policies better than department advisors.

- *Need a signature?* If you need something signed and your advisor continues to break appointments with you, contact the departmental secretary and explain the situation. Secretaries, used to dealing with that particular professor, are usually very helpful.

Can't Stand the Professor

133

When the Going
Gets Tough . . .
Dealing with Ruts
and Unmarked
Pitfalls

Ego too big for the room? Clearly more interested in research than teaching? Try not to let it get to you. Consider it practice in dealing with all the frustrating people you will encounter in life. It is not a good idea to take an individual complaint about a professor (or advisor) to a dean or department chair because rarely (and sadly) do students have any input into the tenuring of professors. Complaining in this forum will only make *you* look like a bad seed.[1]

Poor Professor Solutions

- *Deal.* Don't take it personally.
- *Switch sections.* Find a different professor and give as your reason: "time conflict."
- *Do the best you can.* Why make it worse by not learning anything?
- *Try to work with the TA instead of the professor.* This applies to homework help.
- *Is it the professor or the course? Consider this:* Is it the professor you really dislike or the material he or she is teaching? If it is the material, check whether there is a substitute course. If it is the professor, try not to hold it against the material.

Girlfriends, Boyfriends, and Other Possibly Neglected Distractions

If they are not in engineering, they probably won't understand why you spend much more time with your books and calculator than with them. Just hope they have a life apart from waiting for you.

[1]If you have a complaint where your silence could continue to negatively affect you and/or others, you *should* report it. However, it would be to your advantage to take nondepartmental routes such as a school counselor, RA (resident advisor), or university ombudsperson.

CHAPTER 10

Balancing It All

The ideal engineer is a composite . . . [S]he is not a scientist, [s]he is not a mathematician, [s]he is not a sociologist or a writer; but [s]he may use the knowledge and techniques of any or all of these disciplines in solving engineering problems.

—N. W. Dougherty (1955–)
Professor of Civil Engineering at the University of Tennessee

There are two mountains to summit while passing through our years of engineering education. The first, a balancing act—*life, school, and sanity*—must be maintained, and the second—*the Ben Balance* (as in Ben Franklin)—must be achieved. The sooner we are aware of our need of balance and the sooner we are able to achieve and maintain it, the less stressed and more knowledgeable we become. It is possible to have a fulfilling social life while still excelling in our classes.

LIFE, SCHOOL, AND SANITY

It takes most engineering students one to two years to find their life, school, and sanity (LSS) balance. Freshman engineering students usually take one of three routes:

1. You overwork, overstudy, don't get out much, but get grades to be proud of. (No life, school, borderline sanity will cross over to the dark side your sophomore year.)
2. You absorb college life to the fullest by going to every football game, keg, and Greek row party, but college is tougher than high school and your grades reflect this. (Life, no school, sanity is questionable.)
3. You have a social life, pretty good grades, and enough stress that implosion is a very real possibility. (Life, school, far from sanity.)

Catching on? The goal, of course, is to have and maintain a balance of all three if you weren't lucky enough to be born with the quality of adaptable equilibrium. This isn't to say that life, school, and sanity are all equal. They aren't and they shouldn't be. Picture three glasses of water (of various sizes if you like) on a tray. You are trying to carry the tray with the glasses through an obstacle course without spilling a drop. Keep the tray balanced and you will spill none.

Don't take years to find your LSS balance—start now! Each of us has an individual system of optimum organization, so a balancing act for one person may or may not apply to another. Think! A good starting place is to consider a time in your life when everything seemed to be going *perfectly*. What were you doing that made things run smoothly? Here are more ideas to get you started on finding your LSS balance:

- *Stay healthy!* Read the next section.

- *Fight ruts.* A bad rut can mean no life, no school, and no sanity. See Chapter 9 for strategies to pull yourself out of the most common ruts.

- *Give yourself a break.* Beating yourself up for a bad grade (or even a bad relationship) does no good. Use the negative energy for extra studying, a four-mile run, or something else productive. Doing poorly on a test can occur even when you know the material inside out. It's just the nature of engineering tests.

- *Get organized.* Know what is due when. Keep track of assignments, meetings, and social events. Choose your personal assistant: lists, notebook, day planner, sticky notes on the bulletin board, wall calendar, or Palm Pilot.

- *Get involved!* No matter how much time you think you don't have, everyone has time for at least one extracurricular activity. A real college education is about experiencing everything that interests you. Involvement with diverse school clubs and organizations also ensures that you have a good balance of friends outside of engineering.

KEEPING YOUR BATTERY CHARGED AND MOTOR RUNNING . . .

Plentus venter non studet libenter
(A full belly does not study willingly)
—Latin Proverb

The easiest way to keep your LSS balance is to stay healthy. You've undoubtedly heard it before. Freshman orientation week tends to include a barrage of information on the dangers of drinking, lack of sleep, and the importance of nutrition and exercise. Let's look at the biggies that most affect engineers.

Caffeine: Engineer's Friend or Foe?

It's 10:00 P.M., and you have a final exam the next morning. You are exhausted after having two exams already today. You estimate that you have a minimum of 4 hours of studying to do before you feel like it's worth even showing up at the exam. You need to wake up and start studying. What's the natural tendency? Plug in the coffee machine. Put a six-pack of soda in the freezer for rapid cool. Hunt down some of those stay-awake pills. Is that a good idea? Probably not.

Know the caffeinated facts:

1. *Caffeine can stress you out.* Caffeine is a stimulant that mimics the effects of adrenaline, the principal neurotransmitter of a stress response. In moderate amounts (50–300 mg), caffeine acts as a mild stimulant by increasing the heart rate and blood pressure. Excessive amounts (>400 mg) can cause anxiety, insomnia, headaches, jitters, nausea, and even irregular heartbeats.

Where Those mg Come From:
Know How Much Caffeine You Are Consuming

Coffee (5 fluid oz.)	mg
Drip (auto)	137
Drip (nonauto)	124
Percolated (auto)	117
Percolated (nonauto)	108
Instant	60
Decaffeinated	3

Tea (5 fluid oz.)	mg
Imported	54
U.S. brand	46
Oolong	40
Green	31
Instant	28
Decaf	~1

(All applicable if brewed for 5 min.)

Soda (12 fluid oz.)	mg
Mountain Dew	54
Coke	45
Pepsi	38
RC	36
7-Up, Sprite	0
Fresca	0
Hines Root Beer	0

Chocolate	mg
Baking chocolate (1 oz.)	25
Sweet dark chocolate (1 oz.)	20
Milk chocolate (1 oz.)	6
Chocolate milk (8 fl. oz.)	5
Hot chocolate (6 fl. oz.)	5

continued

Nonprescription drugs (standard dose)	mg
Alertness tablets	200
Diuretics	167
PMS relief pills	60
Cold/allergy medicine	22–36
(Some) pain relief pills	20

Source: For nonprescription drugs: Labels. Everything else: Pennington JAT, Church HN; *Bowes and Church's Food Values of Portions Commonly Used,* 14 ed. (Philadelphia: Lippincott, 1995) through the fab Healthy Devil website at Duke University.

2. *Caffeine is a diuretic.* Let's leave it at that.
3. *Caffeine is a drug.* Regular caffeine drinkers or users can suffer withdrawal symptoms (even if they miss one day!), which may include headaches, irritability, and mild depression.
4. The rest of it probably ain't so good for you either . . .

   - *Carbonated drinks* are high in phosphorus, which depletes calcium in your bones.

   - *Tea and coffee* contain tannin, which, when consumed within an hour before to an hour after mealtime, significantly reduces iron absorption.

   - *Diet sodas* have aspar-what? The jury is still out on aspartame, the sugar substitute in diet sodas. There has been much debate on possible health risks. The Center for Science in the Public Interest recommends that "if you consume more than a couple of servings a day, consider cutting back. And to be on the safe side, don't give aspartame to infants."

5. *There are ways to kick the habit!* Switch to decaf. Try herbal teas (they're caffeine free). Try mixing half decaf with half regular. Read labels. Those who have kicked the caffeine habit say that their study habits have improved because they can focus better.

Is It Possible to Spend Too Much Time with Your Computer?

The average American college student uses a computer for 2.9 hours per day. That's almost a whole day and night each week!

You wouldn't think so. As an engineer, however, the postcomputer marathon feeling is familiar: stiff shoulders, strained eyes, sluggish fingers, and a dull ache in the lower back. When sitting for a long period of time, whether studying *or* hacking, one all-important thing you should do that is guaranteed to improve your cranky mood and muscles considerably is:

Take a 5-minute break every hour (or less).

Stretch. Walk around. Harass your roommate. Splash cold water on your face. Just move! Don't wait until you have pain—be proactive! What else can you do? Keep reading.

Watch Your Eyes!
Problems have been associated with too much time looking at a computer monitor. Farsighted users may experience blurred vision and discomfort and those

with astigmatism may experience eye soreness and headaches if looking at a display for an extended period of time.

To fight eye fatigue:

 Clean your screen regularly.

 Maintain a high contrast between the text and the background. No green on red!

 Try not to work in the dark. Your roommate won't mind the desk lamp.

 Adjust your screen to your viewing pleasure. When you are working, your eyes should line up vertically at the center (or slightly above the center), between 18 and 28 inches away (assuming 20/20 vision) from the screen.

 Crank up the zoom. Just because you'll print the paper in 12-point font, doesn't mean you need to read it on screen that size. And you don't need to be over 65 to use the extra large icons or buttons.

 Invest in a large monitor. Overall monitor size is measured diagonally in inches, however, this dimension does not necessarily describe the viewable length, which may be as much as 2 inches less than the given size. A 17-inch monitor or larger is recommended.

Is Your Monitor Eye and Brain Friendly?
Two nerdy things to take note of: *refresh rates and flat screens.*
- An image is redrawn onscreen several times a second (think about when you see an undoctored computer screen on television); this is the flicker or *refresh rate.* Although folks vary in their sensitivity to flicker, a refresh rate in the mid 70s (Hz) should be annoyance free to all users. Ergonomics researchers, however, suggest that a refresh over 100 Hz would be best, but this could reduce the efficiency of your frame.
- Monitors with *flat screens* are more brain efficient than traditional curved screens. The reason is this: the curvature deflects light into your eye, creating glare, and introduces a slight distortion. You will likely not notice these effects because your brain is working overtime just to compensate.

Okay, so what does this mean since you aren't going to buy a new monitor? Be selective about which computer you choose when you go to the school computer labs, fiddle with lighting to reduce glare, and don't be timid about experimenting with the monitor's control panel.

Researchers of keyboard usage found that computer users tend to use five times the amount of force needed to activate a key.

 Go pro. Buy a copy holder that attaches to the side of your monitor to hold pages. This will eliminate continuous refocusing of the eyes and frequent neck movements between paper and screen.

How Are Your Wrists Holding Up?
Are you a CAD jockey? Do you engage in debates that extol the virtues of the three-buttoned mouse? Do you brag about how many lines of code you can write in a night (whereas another major might brag how many pages he or she can turn

out in a night)? In the past decade, repetitive motion disorders have increased alarmingly with many of them attributed to the computer. The National Institute of Occupational Health determined that eight or nine repetitive movements in a minute did not provide adequate time for the wrist to produce enough lubrication for its narrow passageway of ligament and bone (the carpal tunnel).

To stay out of the carpal tunnel:

- Move your whole arm instead of just your wrist when doing mouse work.
- Pick up an ergonomic keyboard. A wireless ergonomic keyboard and mouse set are even better (not to mention cool, too).
- Keep a loose, relaxed grip on the mouse. A clenched grip means you should probably take a break anyway.
- The optimum angle for your forearms and hands when typing is approximately 12° below the horizon. This often works out naturally if you place the keyboard on your lap.
- Wrist pads are good support when you *aren't* typing.

Sit Smart (at Home)

After reading Chapter 5, you know how to sit smart in class, but what about at home? Some companies actually have lunchtime seminars on how to sit "correctly." The ideal chair for long hours of sitting at a desk is one that is adjustable and provides firm, comfortable support. Doesn't sound much like the standard-issue dorm desk chair, eh?

How to prep your chair to work with you:

- Do a quick upholstery job: Put a pillow on the seat to get maximum comfort!
- Adjust the floor with the college edition dictionary and thesaurus so that your feet rest flat.
- Check out the back of the chair, too. Add another pillow, if necessary, to make sure it supports your lower back. If this doesn't work, you can find lumbar pillows and other higher tech items online.

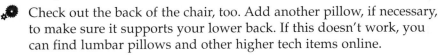

THE "HOMER"

THE "NOTRE DAME"

THE "GANDHI"

With apologies to Matt Groening

Fidgeting is good for you! Humans are comfortable in over 1,000 resting places or postures, but never staying in any one position for very long. Fidgeting is your body's natural mechanism for redistributing stress to other bones and soft tissue and giving your muscles a rest.

 Insure that your monitor, keyboard table, and chair line up so you don't have your head or body angled while you work. Putting the monitor in the corner of the desk may give extra space on your desk—but it may be a pain in the neck (literally!)

How is your typing? If you are a hen-pecker or need to stare at the keyboard while typing, be warned, you will eventually have neck and/or back problems. Does this mean you need to take a typing course? No, just look online to get the basics (which fingers go where) and force yourself to stick with it. People who have changed their typing style cold turkey say that the first few days are frustrating, but after that they are cruising.

Anxiety and Stress

College students feel the pressures of anxiety and stress not only in class, but also in regular day-to-day life. Both are brought on by transition and change. Considering the average college student moves six to eight times in four years—that's not much time to settle into a routine. Reentry students may not be moving as often, but they feel the stress of trying to balance their family and/or job with classes.

Anxiety
Anxiety is the apprehension you feel in response to a real or perceived threat. Reactions are brought on by a sense of distress or through the use of drugs, such as cocaine, caffeine, alcohol, and amphetamines, that affect the nervous system. Reactions can range from mild uneasiness to intense panic. You may have heard people say that they had an anxiety attack during an exam and simply couldn't function. It can be a frightening experience.

 Physical signs. Dry mouth, flushing of face and neck, nausea, lightheadedness, hyperventilation, increased heartbeat, sweating, tremors, and muscle tension.

 Psychological signs. Apprehension, difficulty sleeping, restlessness, irritability, fear, panic, impatience, lack of concentration.

Stress
Stress is the response of your body when your internal balance is disrupted (by positive, negative, unusual, or even normal events) and your system must readjust. Stress is a very common aspect of everyday life and can be beneficial in situations such as taking an exam or driving in a snowstorm where heightened alertness is desirable. Stress becomes a problem when it exceeds a productive level and interferes with your ability to work effectively. Unmanaged high levels of stress can cause physical deterioration and illness after a period of time.

- *Physical signs.* Chronic fatigue; change in appetite; increased use of drugs, alcohol, or nicotine; aches and pains for no reason; change in sleeping patterns.
- *Psychological signs.* Irritability, changed behavior, difficulty focusing.

Prevention Is Key

- *Exercise and eat well.* Mom was right. Exercise and eating well increase your energy level. Twenty minutes of exercise three times a week. Three balanced meals a day.

> ⚠️ **What you consume (or don't consume) can affect your memory.** Everyone handles stress differently. During exam week you may forget to eat or you may eat too much. Maybe you celebrate one exam down with a few down the hatch. Rediscover the food pyramid from junior high health class! Preliminary research has shown that crash dieting or overeating can negatively affect how much you learn and retain. Additionally, frequent or heavy binge drinking has been linked to the death of brain cells. Don't waste your tuition!

- *Get sleep!* The average person needs 8 hours. Aim for 6 hours (at least). Getting enough sleep allows your mind to unwind (psychologists believe that dreams can be an outlet for anxious and stressful feelings) and your body to repair itself. Tasks that require thinking (e.g., homework) can take much longer than usual if you are sleep deprived because it is much harder to focus. It matters not only how

much you sleep, but also when you sleep. Consecutive sleeping allows you to cycle through all the necessary sleep stages so that you feel rested.

> **The All-Nighter: Sometimes It Is Unavoidable . . . But Try to Avoid It!**
> Two things work against you when you pull an all-nighter. The first is that when you are tired, your body has difficulty focusing on the material you are attempting to learn. It will take longer for you to remember, and you may not even remember the information very well. Your second obstacle is that you are "cramming"; you are attempting to learn by studying for long uninterrupted intervals. Psychologists call this "massed practice" and have found that it is much less effective than "distributed practice" where you spread the studying out over many days or weeks.

 Find ways to relax. What works for you? Classical music in a dark room? A warm tub and a good book? A quiet half hour checking out the evening view from a hilltop? A morning jog before classes?

 Hang out with people who give off good vibes.

Ask for help when you need it. And *don't* be embarrassed about it. All schools have a counseling center that offers advice on stress management techniques like muscle relaxation and meditation. They also can provide professional help on an individual level.

EXTRACURRICULAR ACTIVITIES

School clubs and organizations are cheap, fun, and mind expanding (in the best way), and they offer a good change from the schoolbooks. Engineering students tend to steer clear of getting "too involved" on campus because they don't think they have the time. True, you may not have the time other majors have, but don't sell yourself short! Many students find they complete the same amount of schoolwork whether they have several outside commitments or few. With less to do, engineers take more time to complete assignments than is necessary. It's a form of Parkinson's Law (with some literary license): Work expands to fill the time allotted.

Usually at the start of every school year there is a campus organizations fair on one of the school's big green lawns. You don't have to be a freshman to ask freshman questions and try something new. Organizations and clubs are a great opportunity to dabble in the areas that interest you.

Try to do most of your sampling your freshman year. The first year you have a bit more time and you won't be tied down with too many responsibilities. As your schoolwork intensifies with (sophomore and junior) years, you should pick your favorite activities and stick with them.

Look into popular campus activities even if they might not be your cup of tea. There are reasons why they are popular.

Involvement doesn't have to conflict with study time. Go on an alternative spring break program (students who do community service) or, after your freshman year, become a freshmen orientation leader.

 Excuse yourself, with apologies, from (minor) events. These can interfere with work or are just seemingly pointless.

 Involvement produces good leaders.

THE LIFE OF AN ENGINEERING STUDENT—WORK HARD, PLAY HARD

Alas! Something has to give to do everything, right? Well, it depends on how you define *everything.* Engineers don't have the freedom (like some of our eloquent nameless-major friends) to spontaneously decide to have margaritas, play glow-in-the-dark frisbee golf, and then serenade the early morning shoppers at the convenience store across the street from campus. But then, *that* sort of activity can be saved for Christmas break with your high school friends. However, since a few hours of focused work can crank out a problem set, tell your friends that you are skipping the preparty, but you'll meet them later. You may have to give up some of the fun, but not all.

 Be fashionably late to parties. You can study longer (especially if your roommates go out) and you don't usually miss anything anyway.

 Don't feel like a nerd for staying in on a weekend night to get some work done. Staying in on Saturday night can be a good trade for an outing on Thursday night.

 Do blow off your work every once in a blue moon. How often do blue moons come? This is OK for special occasions.

 Think before you drink. If you don't get any work done with a hangover, then don't get blasted Saturday night if you plan on cramming all day Sunday for a test at 9:00 A.M. Monday.

 Plan ahead for weekend events. Don't miss the road trip to Mardi Gras or Graceland to do work, unless you really have to. Get as much completed as you can before you leave. Don't take your books with you. Have a good time. Those memories last longer than a mediocre grade on a Ceramics quiz.

Approximately one-half of all undergrads receive some form of financial aid.

PLAYING A VARSITY SPORT OR WORKING WHILE IN SCHOOL

Playing a varsity sport while in school full time is very much like holding down a part-time job. And whether you work 5 hours or 20 hours a week, it can be stressful to find time to do everything you would like. If you are an engineer committing more than 10 hours a week to a sport or job, free time and social activities will become a rare luxury. The difficulties are more than not having enough time to study and to complete assignments; there also are time conflicts such as work or training schedules that conflict with laboratory classes, review sessions, and TA and professor office hours. Students who are able to successfully balance school with a sport or work usually do so in one of two ways:

1. From the beginning of their freshman year, they successfully integrate sports or jobs into their school schedule.
2. After getting a good grounding freshman year and advice from busy upperclassmen, they allot time successfully for school and work commitments.

The above scenarios are completely different in that folks who fall in category 1 might have to work or play a sport to afford to attend school and live, whereas in category 2 the student may have more financial flexibility.

Let's talk about varsity sports and engineering first. You can do both, but it will be tough, but you've already had some training, right? You managed high school playing a sport—an experience that shouldn't be underestimated—and are probably an old pro at time management. College sports is different, though—venues are farther, training schedules are more rigorous, expectations seem to be higher all around. So what can you do to peak in both commitments?

 Keep your TA and/or professor in the loop (especially if you have scheduling conflicts). They will probably make special arrangements with you and be much more willing to help by e-mail.

 Think ahead. Consider what you want to get out of engineering, and what you want to get out of your sport. Almost all engineering athletes say that compromise happens.

Heed the advice of the old timers. Talk to upperclassmen who are athletes.

If you are interested in finding a job for a bit of cash or work-study (financial aid may require it) while in school, picking the right one could give you some good engineering experience and allow you to study on the job.

Job Prospects for Engineering Majors

Below is a short list of suggested jobs (in no particular order):

1. Research assistantship.

Busy, not always gratifying work, medium schedule flexibility (might be able to swing time off during exams).

Unfortunately, undergrads working for professors do a lot of grunt work like cleaning samples, filing and cataloging articles, and helping grad students with data collection. However, the benefits far outweigh any of the mindless work: You get good hands-on experience in a research environment that may even turn into a real research position as an upperclassman; you become acquainted on a personal level with the professor, grad students, and other folks around the department; it looks good on your resume; and frequently you are granted time off during exams. Turn to Chapter 8 for more details and information.

As students, U.S. president Lyndon Johnson worked as a janitor and Nobel economist Milton Friedman waited tables.

2. Computer monitor.

Sometimes very busy, sometimes dead, medium schedule flexibility (often expected to work through exams).

Working at the Help Desk forces you to learn more about computers and the newest software. This is always an asset to the engineer. Some computer labs are very social while others are quiet and tense.

3. Bartender, cocktail server, runner at a nice restaurant or happening bar.

Very busy, little flexibility (unless you can trade shifts, but restaurant/bar managers can grow impatient with students).

So why would you take a restaurant or bar job? Jobs like this pay cash for just a few hours of work if you pick an upscale restaurant or bar. Food runners and cocktail servers are known for being able to work a 4- or 5-hour shift and pick up ~$100. The downside is that you will probably have to give up a weekend night, the income is not guaranteed, and restaurants and bars don't take Christmas and spring breaks.

4. Engineering internship/co-op.

Real work, flexibility varies.

Internships and co-ops provide incredible insight into and experience in engineering as a profession. They can give you direction and an education that cannot be gained by taking classes. The good news is that internships often pay very well, but the drawback is that a good internship can be difficult to balance while attending school full time. For more info, refer to Chapter 8.

5. Checking ID cards or working the reserve desk at a quiet library.

Not usually busy, medium schedule flexibility (usually have to work during exams).

Good for studying, relaxingly calm (but tedious?).

6. An outside interest job.

Busy, but fun; variable flexibility.

Haven't had the opportunity to unzip the legs from your convertible hiking shorts-pants in a while? If you are the hiking, outdoorsy type, see if the rec center has a job in its outdoor education department. Why not get paid for what you might be doing in your spare time? Some of the best-paying, flexible jobs can be found by checking the school newspaper want-ads. Often the psychology department is looking for human guinea pigs at $25 an hour to fill out questionnaires (sometimes worse). You might also see excellent short-term, paying jobs for engineering students needed to test websites or work on textbook solutions manuals.

THE BEN BALANCE

Ben! The same legendary Mr. Franklin exulted in U.S. history textbooks, legendary for the key-on-the-kite-during-lightning picture: inventor, printer, writer, politician, diplomat, and scientist. He was a colonial American Renaissance man with only two years of formal education under his belt. Had he gone to a modern-day engineering school, he might not have come out so well rounded.

Part of the problem is *how* many engineering schools (and thus engineering students) tend to view the nontech classes. The few liberal arts classes an engineering student needs to take are often seen as a *requirement*. And while both Anatomy and Technical Writing may be requirements for a biomed engineering student, Anatomy isn't usually regarded with the same disdain as Tech Writing. Of course, you may be more interested in Anatomy, but Technical Writing is a vital part of the engineering picture, too.

Why do we hear people remark that "engineers must be smart, but—they lack common sense!" Wait! That's not right. Engineers are supposed to be the most logical of the homo sapiens, although . . . well, often their logic is discrete on the infinitesimal scale, and they sometimes miss the macroscopic. Engineering education can sometimes be too one sided and left-brain heavy. A better balanced brain makes for a better problem-solving engineer and a more knowledgeable person.

Where might engineers need some right-brain remedial? How about in the realm of culture and nonscientific current events? They are interested, but they are mostly up to their eyeballs in work and extracurricular commitments. You've heard that education isn't solely in the classroom. That's true, and it's up to you to fill in the gaps and grow like yogurt under a heater. Your engineering classes can't do everything.

So what do you do? Think critically about yourself. Picture the type of person you want to be when you graduate. Pinpoint areas where you think you are lacking and find a fun way to tackle them.

Look where you are lacking and fill in the gaps!

Art

Does the word *art* make you uncomfortable? Do you cringe at the thought of having to draw a self-portrait? Not only is art distantly related to engineering (*technes*, the root of technology, is Greek for *art!*), but it is a necessary tool for the engineer. Think how often you have to draw diagrams when solving problems. Decent drawing skills are necessary for practicing engineers to communicate exactly what they are working on. Getting a little art in your life doesn't mean you have to sign up for a theater or art class—although that would definitely make you a Renaissance person.

The root pencillis of pencil is Latin for brush.

Do the following to revive and utilize your latent artistic talents:

- *If your school has an art library, go poke your head in.* If you think *you* are uncomfortable in an *art* library, try to get an art major to go to the engineering library.

- *Notice the random art exhibits in the student union.*

- *Keep a design log.* Engineers involved in design wouldn't live without them. What products peeve you? Is it a pain to adjust your rear-view mirror accurately ? If you were a design engineer, how would you redesign it? Sketch your ideas and add explanatory notes. Interesting articles and tidbits also can be taped or jotted down in logs. Famous inventions are born in design logs.

- *Read the plaques on the campus statues.* Try to figure out the material and method used to produce the artwork. How is it manufactured?

Go to school plays and concerts, especially if you have a friend participating.

> ☞ **Get the most out of your school!** Go to traveling art exhibits, free speakers, webpage or cooking workshops, goofy dorm field trips, sporting events, cheap movies, theater productions, fly-fishing classes at the rec center, photography or bartending classes at the student center. It's inexpensive, fun, and a good break from your regular schedule.

Out of Touch

Your parents called to ask if everything was okay after they saw on the news that three tornados had touched down within 100 miles of your school. Tornados? It rained a lot and . . . hey, now that I think of it, there were a lot of branches on the ground. . . .

It's easy to lose yourself in your work and miss the boat on current events. The worst part is that the full realization that you have no idea what is going on becomes painfully obvious when you try to keep up a conversation with your newly acquired significant other's father or a professor you would like to impress. The microcosm syndrome can leave you feeling clueless and self-absorbed. Having an idea about what is going on off campus will give you a better grounding and more chances to challenge your own ideas and beliefs.

Be omniscient!

 Pick up a newspaper with your morning soda. Even the comics are a start (you have to glance at the headlines to get there).

 Listen to AM talk-radio during your shower or the commute.

 Get involved with a community service group. The hardest thing about doing community service is taking that first step to get involved. After that, it's easy and fun, and the rewards come back to you tenfold.

> **Do Community Service and Engineering—and get credits toward your degree!** The EPICS (Engineering Projects in Community Service) program, founded at Purdue University in 1995, is now in action at over a dozen universities. Teams composed of freshman through senior students work with local not-for-profit organizations to solve engineering-based problems in their community. Some of the projects students have worked on include a secure distributed database for multiple homelessness-prevention agencies to track and coordinate services, and multisensory doll houses to give children with physical disabilities an environment for active play. Students and community organizations who have taken part in EPICS have given the program enthusiastic thumbs up. If this sounds like something you are interested in, ask your advisor about it or check out the EPICS program's national website:
>
> **epics.ecn.purdue.edu/**

Find an offbeat coffeehouse or bookstore. These sometimes have newspapers from around the world. The fewer people you recognize from school, the better.

☞ **If your university is in or near a decent-sized city, get out and explore!** Seek out local favorites on your own and by asking around. Check the student union and admissions/orientation office for touristy pamphlets and maps. If you aren't in a major city, check the map and find a friend with a car. It's time for a road trip! A weekend or even an overnight road trip to a ball game, the beach, or over the border will inject some fun into a dull semester.

 Seek out extracurricular activities that involve dialogue with others. This could inspire you to think about your own values. Good examples for contacts are environmental organizations, political action groups, and committees that organize campus speaking events.

Life 101

Do you feel left out of political discussions? Do interest rates (and the big deal over them) leave you baffled? You should pick up a myriad of useful life tools before you graduate to make you a better engineer and world citizen, like:

 Intro classes in social sciences and humanities. These are great for a crash course in the human race. Everyone should definitely have an Introductory Economics class to learn all the stuffy terminology and understand annual reports. Other good nonrequired intro classes for engineers are Psychology, Accounting, Public Speaking,[1] and English Composition.

 Go overseas for a quarter. University is one of the few times you will be able to simply pick up your life and move somewhere on the other side of the planet (in a somewhat organized manner) for 3 months. Even if your school doesn't have a program in your choice destination—other schools might. Don't forget to check out financial aid and scholarship options; many organizations like Rotary International offer "Ambassadorial Scholarships."

 Workshops sponsored by your dorm, student government, student center, or career center. Such workshops can teach you the ins and outs of buying (and repairing) a car, how to find the right life/car/fire/accident insurance, and the basics of mortgages and investments. Workshops and seminars are usually quick and useful.

 Opportunity. Just dive in!

The average engineering student takes 11.4 semester units in social sciences and psychology. This breaks down to approximately 8 percent of classes.

---

[1]Some engineering schools will not give humanities/social sciences credit for public speaking. Check with your advisor to see where it figures into your degree requirements and electives.

*L*ooking back over the past 48 years as a civil and professional engineer, I found my 4 years of undergraduate education to have been key to a lifetime of continuing education. Every step taken toward success has demanded a desire to learn about such things as people, philosophy, business, management, legality, methods, and money as well as about the various aspects of engineering. Study never ends! Engineering school provides an opportunity to obtain certain keys that will be needed through a career, but they will not unlock doors without continued interest in learning.

(J.E.C., Civil Engineering (Structural) '52, University of Connecticut)

The Big Picture

So you are learning all this useful, tangible knowledge about the interaction of complicated scientific and computational systems. But how much do you really know about the relation of humans to the technology that engineers generate? Well, let's see . . . engineering ethics might have popped up your freshman or sophomore year in an intro engineering class, but often that's it! Engineering has a colorful history filled with eccentric and fabulously odd people, but unless you read the unassigned historical section in the first chapter of a course text, you miss out. Did you know that the young Isaac Newton carved his name on almost every school bench he occupied? Or that Ada Byron Lovelace, the first computer programmer, was an incorrigible gambler? It's also useful to know how engineering reaches into other realms like law, medicine, philosophy, and sociology.

Learn Engineering from a Different Angle

- *Good classes.* History of Science or History of Technology, Engineering (or Science) Law, Technology and Public Policy, Management of Technology, Technology and Society.

- *Go to campus forums like* "Science & (Pick one): Ethics, Religion, Third-World Countries, Society, Science Fiction."

- *Check out a graduate engineering seminar.* You don't have to understand all of it to get a sense of the cool things being researched and studied in depth—all of which are based on engineering fundamentals.

- *Audit a class for a day.* Granted, having the time would be a luxury, but if a friend's prelaw or engineering class is having a professor come to speak about applying for a patent—and that sounds interesting to you—don't be bashful. Chances are the professor won't even notice your presence, and in any case would probably be flattered by your interest.

- *Do an engineering co-op.* Co-ops are internships for which you get school credit, often with a required paper or sometimes a weekly discussion group. Co-ops allow you to see how engineering and your skills fit into industry. Also, you can earn good money. Turn to Chapter 8 for more info.

- *Field trips!* Take advantage of field trips sponsored by engineering societies. Plant tours are the most common destination. Industry field trips may give you some insight into where you may (or may not) want to work.

Are you an ME who feels lost when you look under the hood of a car? Have you had a circuits class, but wouldn't know a capacitor from a resistor unless someone drew their symbols? You are certainly not alone. One of the complaints heard most often about engineering education is that there is not enough application of taught theory. Part of the reason for this is that many of the professors who are teaching you to become a practicing engineer have *never* practiced engineering "in the trenches" themselves. Wild, eh? Some professors took the academic route of getting a bunch of degrees and then taking a teaching post at a university. So, don't be surprised if you feel you are not getting enough hands-on experience.

Tips about how to get handy:

 Take the offered machine shop class. Even if you are a biomedical engineer, it is still useful to be able to produce your own prototype with your own hands.

 Load up on lab classes.

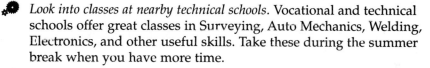 *Read scholarly and popular journals. Scientific American, Discovery,* or any magazine published by an engineering society would be excellent choices.

 Look into classes at nearby technical schools. Vocational and technical schools offer great classes in Surveying, Auto Mechanics, Welding, Electronics, and other useful skills. Take these during the summer break when you have more time.

 Keep your eyes and ears open for on-campus workshops. Especially workshops that teach crash courses on UNIX or how to use multimedia equipment.

Plan your own fun field trip. How about going to an air show or hands-on science museum?

Get involved with competitions put on by professional societies for students. Solar Car Project, Concrete Canoe Race, Aerial Robotics Competition, Autonomous Ground Competition, Aerodesign Competition, just to name a few.

Communication Skills

Communication skills are what many practicing engineers wish they had mastered while they were in school. On-the-job engineers find themselves giving presentations to their co-workers and superiors, writing letters and memos, and constantly interacting with folks of various backgrounds who may or may not have technical know-how. Good communication skills can be acquired quite easily, but you may have to take your own first step because many curricula place little or no emphasis on it.

Communicate with this:

 (Useful) communication classes. Any one (or more!) of the following classes: Public Speaking, (easy) English, Technical Writing, Technology Management, and so forth.

Book clubs and discussion groups. Check with the university bookstore or library.

Check out or purchase a vocabulary book. Some are actually kind of informative—know what *callipygian*[2] means?

Work for the school newspaper or literary/humor magazine.

Any extracurricular group that requires verbal idea exchanges.

Other People

Nope. Not kidding. You needn't look forward to the vacation at home to be around "normal people." As awesome as we engineers are, we do get on our own nerves. Part of having a good balance for ourselves is having a good balance of diverse friends.

More Friends, More Fun: Places to Look

- Community or public service (on campus or off)
- Theme dorms (like French or philosophy)
- Intramurals and off-campus athletic leagues
- Greek organizations
- Religious and political groups
- Common-interest clubs like photography or hiking
- Community organizations

THE COUNTER-CREATIVITY MYTH

You may have noticed how people who place themselves on the science/math end of the spectrum tend to make disclaimers for their art ability, correctness in spelling, and creativity in thinking. Engineers (and not just students) make many disclaimers.

It brings to mind the reported classroom research about singing, dancing, and drawing told by Adam Werbach (who at age 23 was elected president of the Sierra Club, an American nonprofit environmental and public interest organization). As he tells it, a small group of researchers with clipboards in hand visited a kindergarten class. The first question they asked was:

"How many of you can sing?"
The kindergartners (somewhat shyly) all raised their hands; they had just sung "Puff the Magic Dragon."
The researchers then asked, "Well, how many of you can dance?" There were some smiles and all the students raised their hands.
"Good, then how many of you can draw?" the researchers asked. All the giggling children raised their hands.

[2]Having well-shaped buttocks.

The same researchers visited a senior-level class at a university in upstate New York. Again they asked questions.

"How many of you can sing?" The college students looked blankly at the three researchers. A lone hand timidly raised in the rear of the auditorium.

"Uh-huh, and how many would say that they could dance?" There were some smiles, but no one raised a hand.

Lastly, the researchers again asked: "How many people can draw?" A mere six or seven hands went up.

Hmm . . . What happened?

Some of the most famous scientists and engineers devoted a great deal of time to singing, dancing, and drawing and would you have guessed also weaving, poetry, and other fun things?

And there are many more who shatter the counter-creativity myth. *Engineers are inherently creative because they are problem solvers.* Just think of Mac-Gyver.[3] So, with the Ben Balance and LSS Balance you will never emit a disclaimer.

Talk about Well-Rounded!

Who?	How We Would Know Them	Mystery Talent(s)
Sir Humphry Davy	Rubbed two blocks of ice together on a subzero day and watched them melt to study thermodynamics and heat transfer	Poet
Albert Einstein	Theory of relativity ($E=mc^2$)	Musician

continued

---

[3]For those of you who missed the reruns: *MacGyver* was a popular TV show in the '80s (reruns are *still* on) about a guy (named MacGyver) "who solved almost any problem, using only science and wits. He'll use that paperclip to short-circuit a nuclear missile; with chocolate, he'll stop an acid leak." —*TV Guide.*

Who?	How We Would Know Them	Mystery Talent(s)
Galileo Galilei	Greatly improved the telescope, introduced Strength of Materials and Kinematics	Poet
Robert Goddard	Goddard Space Flight Center in Maryland, worked in rocketry, invention of the bazooka	Science fiction writer
Sir William Rowan Hamilton	Law of multiplication of quadrions: $ij = k, jk = i, ki = j, ik = 2j, kj = 2i, ji = 2k,$ $i^2 = j^2 = k^2 = 1$ (used in vector analysis)	Poet
Friedrich August Kekulé	Kekulé structures: graph representation of chemical compounds with abbreviations of elements and dashes	Architect
Johannes Kepler	The three laws of planetary motion	Artist, musician
Ernst Mach	Mach number describing a moving object's speed in air (Mach 1 equals the speed of sound)	Musician
James Maxwell	Maxwell's equations (one of which is Faraday's law) describing electric and magnetic phenomena	Poet, photographer
Siméon-Denis Poisson	Poisson's ratio relating lateral and transverse deformation in a material	Fiction writer
Charles Richet	Developed the first immune serum	Poet, playwright, nonfiction writer
Erwin Schrödinger	Schrödinger wave equation	Weaver

Dear Sir,

> *The fact that I beat a drum has nothing to do with the fact that I do theoretical physics. Theoretical physics is a human endeavor, one of the higher developments of human beings—and this perpetual desire to prove that people who do it are human by showing that they do other things that a few other human beings do (like playing the bongo drums) is insulting to me.*
>
> *I'm human enough to tell you to go to hell.*

—RICHARD FEYNMAN (1918–1988)
American physicist, poet, artist, musician, and safecracker, in response to a request from a Swedish encyclopedia publisher for a copy of the photograph of Feynman (in shirtsleeves gleefully pounding a bongo drum) that adorned each volume of *The Feynman Lectures on Physics.* The publisher wished to "give a human approach to the difficult matter that theoretical physics represents."

CHAPTER 11

Beyond the Bachelor's Degree—Things to Think About Midway Through

Today was good. Today was fun. Tomorrow is another one.
—Dr. Seuss

By the time you are a junior, you should have a general idea of what you want to do after you finish your bachelor's degree. The good news is that even if you don't and you are a senior, engineers tend to have more options than almost any other major (just look around at a career fair). Your analytically trained mind, tech-n-*abilities*, and computer-hacking skills will be a bonus when wandering through the job market.

So why do you have to start thinking about all this now? Because life after college comes up awfully fast when you are a senior. Interviewing for jobs will start in the fall of your senior year. Many graduate school and scholarship applications are due by January, and while you might not have a problem writing essays last minute and then express-mailing them out, getting references from professors always takes a few days. But even before all that, you are forced to consider what you do want to do after you graduate and where you want to do it.

These days graduating engineers can do pretty much anything they want. Philosophy majors can't do a master's in electrical engineering, but electrical engineers can go to business school or law school (and hang out with the philosophy majors). Engineers are fully versatile! With that in mind, these are the three general routes a graduating engineer takes:

1. Become a practicing engineer! (This is the obvious one.)
2. Head to graduate school.
3. Take a route that's not quite engineering. You use your skills in another area.

158

Beyond the
Bachelor's Degree—
Things to Think
About Midway
Through

> ☞ **Some Good (Free) Resources for the Maybe-Job-Hunting, Maybe-Going-to-Grad-School-about-to-Graduate Engineer:**
>
> • *Diversity/Careers in Engineering & Information Technology* publishes two student-focused issues a year that are informative as to what different engineers do and in preparing for the job hunt. You can find this magazine at the dean's office or at job fairs. Their website can be accessed at:
>
> **www.diversitycareers.com**
>
> • *Graduating Engineer and Computer Careers* comes out monthly and covers job-seeking and grad school–seeking activities and issues:
>
> **www.graduatingengineer.com**
>
> • *Google*—no joke. A search on "engineering jobs" or your discipline + "jobs" will turn up a variety if websites such as engineeringjobs.com, engineering.com, mechanicalengineer.com, and civilengineeringjobs.com. Surf around to see what the possibilities are and what is out there!
>
> **www.google.com**

A LICENSE TO ENGINEER?

Before you get out a map to determine which route to take, consider this: No matter which career direction you choose to take—If you live in the United States, take the Fundamentals of Engineering (FE) exam your last year (or earlier if you'd like) of undergrad.

The FE examination (it may be called the EIT or Engineering in Training exam depending on the state you live in) is the first step in becoming an officially licensed engineer. The rules governing licensure vary from state to state, but the process is fairly consistent. It is an 8-hour state board exam that you will take in some random school cafeteria or gymnasium if you aren't lucky enough to have your school host a site. After you have passed the EIT and have worked a few years under a PE-certified engineer, you can apply to take the PE (Professional Engineer) exam that is specific to your discipline. With PE certification you get to put the cool initials, PE, after your name when you sign documents or letters.

The EIT is offered twice a year (April and October). Depending on your state, you may need to have graduated from an ABET-certified university or college (turn to page 48 for more info) to be eligible. Since most students take the exam their senior year, you should start looking into the registration process and all that good stuff by the end of your junior year.

The FE (a.k.a. EIT) Rundown: How the Process Works

1. Find your school announcement bulletin board with informational packets about becoming a certified engineer. If you can't find any information at your school, check out **www.engineeringlicense.com.**
2. If your school has FE packets—grab one. Read it. If your school doesn't seem to have much information, everything you need is at

engineeringlicense.com. Follow the links to FE, then to your state licensing board. Registration contacts and information should be provided there.

159

Beyond the
Bachelor's Degree—
Things to Think
About Midway
Through

3. Fill out all necessary postcards, forms, and so forth, and write a check for the application fee (registration and fees vary from state to state).
4. Mail it!
5. Sign up and attend an EIT prep course (often offered by your school) *if* you think you may need it. If your school doesn't offer a prep course, the NSPE has information on test preparation. If you would rather study on your own, the most comprehensive book is *Fundamentals of Engineering: The Most Effective FE/EIT Review (Fundamentals of Engineering, 10th edition)*, edited by Merle C. Potter. There are cheaper, discipline-specific options also, so check online or at your local bookstore.
6. You will receive a packet in the mail with the site of your exam, a formula and table reference booklet that will be the same as the one you will be given at the test, and some other information.
7. Familiarize yourself with the formula and reference booklet for speedy appropriate formula-retrieval on test day.
8. Take the test.

Nearly 75 percent of people who take the FE exam pass it the first time.

 EIT/FE pass rates vary from year to year and discipline by discipline. You can psych yourself up/out at.

www.ncees.org/exams/pass_rates

The Examination Itself

The EIT or FE exam is an **8**-hour (in case you missed that earlier), multiple-choice exam divided into two 4-hour sessions with a 1-hour break for lunch. The morning session is a general exam (common to all disciplines) with 120 questions worth one point each. Questions cover all kinds of topics: Mathematics (20 percent), Electrical Circuits (10 percent), Statics (10 percent), Chemistry (9 percent), Thermodynamics (9 percent), Dynamics (7 percent), Mechanics of Materials (7 percent), Fluid Mechanics (7 percent), Materials Science/Structure of Matter (7 percent), Computers (6 percent), Engineering Economics (4 percent), and Ethics (4 percent). You can see that mainstream disciplines have an advantage over narrower disciplines, however, all engineering majors cover enough of this material to pass the exam.

A few things to know before exam day:

- You need a 70 out of a maximum score of 100 to pass the exam. This is not a raw score, but some number arrived at through mystery calculations within the NCEES. They believe this method to be fair.

- You are allowed to bring one "noncommunicating, battery-operated, silent, nonprinting" calculator to the exam.

- Units are generally in metric, but English units may show up where appropriate in CE, IE, and ME sections.

Following are some typical questions you can expect on the morning test.

Some Might-Be's on the Morning Exam:

6. The paraboloid 2 units high is formed by rotating $x^2 = y$ about the y-axis.
   Its volume is:
   (A) 2π
   (B) 4π
   (C) 6π
   (D) 8π

23. Consider the following program segment:
```
x = 2
y = 0
z = -1
for i = 0 to 2
(   x = x + 1
    y = x - y
    z = z*y
)
```
   What are the final values for x, y, z?
   (A) $3, -3, 3$
   (B) $5, 2, -3$
   (C) $5, 4, -4$
   (D) $4, 1, -1$

49. In a series RLC network the applied voltage is $v(t) = 56.3\sin(144t)$ and the
   circuit current is $i(t) = 4.7\sin(144t + 0.27)$. The power delivered to the network
   at $t = 3 \times 10^{-3}$ seconds is:
   (A) 71.5 W
   (B) 84.2 W
   (C) 94.7 W
   (D) 140.4 W

73. An isobaric process is one with
   (A) constant volume
   (B) constant pressure
   (C) constant volume
   (D) zero heat transfer

110. Flow through a large river is to be scaled for a movie. The river has an
   average depth of 4m and a flow rate of 95 m^3/s. The pump used for modeling
   the river has the capability of a 0.75m^3/s flow rate. What depth should the
   movie river be scaled to?
   (A) 0.032m
   (B) 0.048m
   (C) 0.47m
   (D) 0.80m

(Answers, in case you were curious: 6. A; 23. D; 49. A; 73. B; 110. D)

The afternoon session for the EIT or FE is offered in five disciplines (chemical, civil, electrical, industrial, or mechanical) and one nonspecific general exam. All exams are similar in format with 60 problems worth two points each. While the afternoon problems require more work than the morning problems, they tend to be grouped so that one answer leads to the next problem, or many problems use the same diagram. Although past scores on the discipline-specific exams have been slightly higher, many students still opt to take the general exam because they are already studying a broad range of problems for the morning session.

161

Beyond the
Bachelor's Degree—
Things to Think
About Midway
Through

Afternoon Problems that Might Make an Appearance:

Given a set of equations represented by $[a_{ij}][x_j] = [y_i]$:
$$3x_1 - 1x_2 = 11$$
$$x_1 + 3x_2 = 17$$

1. What is the adjunct matrix $[a_{ij}]^+$?

(A) $\begin{bmatrix} 1 & 3 \\ 3 & -1 \end{bmatrix}$

(B) $\begin{bmatrix} 3 & 1 \\ -1 & 3 \end{bmatrix}$

(C) $\begin{bmatrix} 3 & -1 \\ 1 & -3 \end{bmatrix}$

(D) $\begin{bmatrix} -3 & 1 \\ 1 & 3 \end{bmatrix}$

2. What is the inverse matrix $[u_{ij}]^{-1}$?

(A) $\begin{bmatrix} -3 & 1 \\ 1 & 3 \end{bmatrix}$

(B) $\frac{1}{6}\begin{bmatrix} -1 & 3 \\ 3 & 1 \end{bmatrix}$

(C) $\frac{1}{9}\begin{bmatrix} 3 & -1 \\ 1 & -3 \end{bmatrix}$

(D) $\frac{1}{10}\begin{bmatrix} 3 & 1 \\ -1 & 3 \end{bmatrix}$

3. When using Cramer's rule, x_1 can be found by evaluating a determinant $|b_{ij}|$ and then dividing by the determinant $|a_{ij}|$. What is $|b_{ij}|$ for x_1?
   (A) 16
   (B) 17
   (C) 33
   (D) 50

4. What are the eigenvalues or characteristic values of the matrix $[a_{ij}]$?
   (A) 1, 2
   (B) 1, 3
   (C) 2, 4
   (D) 3, 3

continued

162

Beyond the
Bachelor's Degree—
Things to Think
About Midway
Through

Consider the following network:

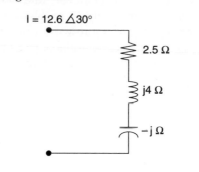

$I = 12.6 \angle 30°$

2.5 Ω

j4 Ω

−j Ω

34. What is the average real power delivered to this network?
(A) 64 W
(B) 112 W
(C) 198 W
(D) 256 W

35. What is the apparent or complex power delivered to this network?
(A) 138 VA
(B) 309 VA
(C) 452 VA
(D) 703 VA

36. What is the power factor for this network?
(A) 0.54
(B) 0.64
(C) 0.66
(D) 0.78

(Answers: 1. B; 2. D; 3. D; 4. C; 34. C; 35. B; 36. B)

It doesn't look too terrible, does it? If you think you need more practice testing, the engineeringlicense.com website has a free downloadable practice FE exam you can take at:

www.ncees.org/exams/study_materials/fe_materials.php

Too many engineering students do not take the EIT, some because they didn't know they should and others somehow justified skipping out. Take it! You have nothing to lose (failing it is the same as not taking it at all). Your EIT certification is the government's recognition that *you are an engineer*. Even if you are heading to business school, take it! You may change your mind about corporate life. And who wants to go back and study for an engineering exam that tests on four years of education (some of which you never even saw)? Still uncertain? See *No Lame Excuses about Taking the EIT Exam!* on the next page.

After putting this final test behind you, you can put your education to the test.

No Lame Excuses about Taking the EIT Exam!

1. *It costs too much.* Ouch! It does cost a little over the non-fun allowance quota. Gotta think of it as the gift that keeps giving.
2. *Don't have time to study.* It's your senior year! Projects are due! Job interviews! You haven't made one review session anyway. People have been known to pass with a good memory and minimal (no) studying. Don't forget a formula book is given to you for the exam.
3. *Don't need it to be a good engineer.* You're right in theory, *but* legally only a licensed engineer can offer services directly to the public. Especially if you are a civil or environmental engineer, with a PE, you will also be a higher-paid, faster-promoted engineer who can prepare, sign, seal, and submit plans and drawings to a public authority for approval.
4. *Eight hours of a weekend is too much to give up at the end of your senior year.* It *is* too much, so either take it in the fall before senioritis sets in or just take it at the end of the year anyway. And after reliving your engineering education compressed into eight hours, justifying going out for an immediate beer is no problem.
5. *It gets harder the longer you are out of school!* The average FE pass rate of 75 percent for engineering students finishing up university drops to a 52 percent pass rate for engineers who have been out of school for two years. (The average pass rate is even lower for engineers who have been out of school longer than two years.)

Engineers with PE licenses earn 15–25 percent higher salaries than their unlicensed counterparts.

ON TO BIGGER AND BETTER... THINGS? PLACES? ADVENTURES?

Where are you heading? Where will you live? How will you pay off your loans? What are you going to do with the rest of your life? The final year of school brings some big (and frequently asked) questions. What can you do with an engineering degree? In brief: a lot. Fresh out of school you are the most employable of all bachelor's degree holders. While you might be focused on taking an engineering job, there are infinite other possibilities, too.

The Engineering Job

Senior year becomes an exercise of a different kind of equilibrium: keeping up with school, arranging interviews (and possibly being flown to them), and outings with friends. An early jump on the job search process can leave you with valued free time and sanity once the schoolwork kicks into high gear. The following things will help you get your first engineering job:

 Having had summer engineering internships. Internships serve multiple purposes. They look great on resumes, help you make decisions about your future (did you like what you did?), and can lead to jobs after you graduate. Turn to Chapter 8 for more info.

 Having taken a co-op or an industry-sponsored class. A co-op works much like an internship except you get school credit. Industry-sponsored classes are often upper-level design or research-type

1 + 1 = ___

The highest median annual income by a major branch in engineering is petroleum engineering at $102,500. Nuclear and sanitary engineering were second and third at $88,050 and $84,900, respectively. (2003)

164

Beyond the
Bachelor's Degree—
Things to Think
About Midway
Through

classes where outside companies or agencies contribute time, money, and possibly facilities for students to work on "real" projects. Strong performance during a co-op or on an industry-sponsored school project often results in a job or internship offer. Employers often look for students who have co-op experience because they believe them to be more committed and mature in their approach to work. More info on co-ops can be found in Chapter 8.

- *Use of your university's career center.* The center will lead you through the maze of resume production, interview strategies, networking, kiss-up cover letters, job fairs, and job searches. It's often a good idea to start attending workshops your junior year so you aren't slammed at the start of your senior year.

- *Keeping your ears open at the department.* Local companies and alums looking to fill jobs or recruit candidates often call professors.

- *Work connections.* Nearly 80 percent of jobs are landed through family, friend, or professional society connections. Something to think about!

- *Keeping it all in perspective.* Don't stress about making a "big life decision." Most college graduates switch from their first job to another within the first three years.

- *Having strong verbal and written communication skills.* Well-developed oral and writing skills are becoming almost as important as strong technical abilities. No matter how you cut it, engineers need to be able to communicate effectively and clearly regarding what they are working on and the proposed solutions. Second-round interviewers are even known to ask students to give impromptu presentations

("That's funny—the overhead projector worked this morning, would you mind going ahead without . . . ") during interviews.

165

Beyond the
Bachelor's Degree—
Things to Think
About Midway
Through

 Portfolios aren't just for art students. Some companies (particularly those in design domains) request that you submit a portfolio or bring one to the interview. What is in a portfolio? A good portfolio has a cross-section of photographs, drawings, and schematics of projects on which you have worked. If you submit a portfolio to be reviewed when you are not present to explain its elements, include short descriptions of individual projects and captions for all visuals. Highlight your multiple talents by not simply sticking to your technical and engineering projects—include a few photographs if you took a photography class. While you do want to tailor it to the company (like your resume), those requesting portfolios are usually looking for evidence of diverse talents.

 Creating an interesting, professional home page. HTML[1] is easy with the aid of a tutorial book, some WYSIWYG[2] software, or a campus workshop. Put your portfolio up on the Web! Movies (of projects!) earn bonus points from recruiters.

The Quick Guide to an Engineering Resume

Resumes now come in three general formats you could *employ* to find employment: (1) the traditional paper resume, (2) the web resume, and (3) the new kid on the block, the at-a-glance resume. The latter two are generally the same thing simply in different mediums, but the at-a-glance resume is a summary of your resume.

The key to a good resume (no matter which form) is the same as giving a speech:

Tailor it to your audience.

Although you might be applying to several jobs in the same field, it is worth considering how well your resume matches each job individually.

A.   The Traditional Resume

The traditional paper resume printed on 8-1/2" × 11" stock is still the standard default. Resumes can follow many formats, so check with your university or college career center for its recommended format. The following format is standard for engineering grads. (See Appendix E for an example.)

Some Rules:
1. Use one side of *one page.*
2. Margins should be no less than 3/4 inch on the right- and left-hand sides and 1/2 inch at the top and bottom of the page. This allows sufficient buffer space for photocopying and binding (for resume books).
3. Try to stick to white paper. Gray and beige paper do not reproduce as well as white.
4. For maximum clarity (especially for faxing and photocopying), use a font no smaller than 10 points.
5. It's better to be simple than slick.

continued

---

[1]Hyper Text Markup Language.
[2]What You See Is What You Get (pronounced: *wizzy wig*).

166

Beyond the
Bachelor's Degree—
Things to Think
About Midway
Through

The Breakdown

1. *Identifying info.* Include your name, permanent and school addresses, e-mail address, home page URL, and phone number with area code.

2. *Job objective.* Make a brief statement about the kind of position you are looking for. Your objective statement will alert the employer to your specialty, interests, and the internal department or division to which to forward your resume. Some good examples:
   - A civil engineering position, especially in environmental planning or geo-technical engineering. (CE)
   - A research position in computational mathematics preferably in close proximity to an apple tree. (Isaac Newton)
   - A position in digital design or signal processing. (EE)

   A side note: in some situations you may want to make your resume as broad as possible, and having a job objective can be restrictive.

3. *Education.*
   a. *The main stuff.* List your school(s), major, degree to be received (B.S., B.E., A.B., etc), projected graduation date, and dates of attendance. Schools should be listed in backward chronological order with most recent at the top. Summer school at home, junior college, and overseas programs also can be included.
   b. *Optional inclusions.* In situations where a transcript has not been requested, it may be appropriate to list which classes you have taken. If you do list classes, only include those that are specialized to your discipline (the recruiter assumes, for example, that you have taken calculus).

4. *Experience.* Experience can be any paid or unpaid jobs or duties that illustrate your excellent leadership, interpersonal, communication, and technical skills. Experiences also should be listed with the most recent first. Include dates of employment, job title, employer name and location, and responsibilities and accomplishments. Start descriptive sentences with an action-oriented verb and highlight activities that would most interest your employer.

5. *Honors/Awards.* List honor societies, awards, scholarships, and distinctions in order of importance. Tailor the order of importance to the reader.

6. *Additional information.* This is the place for any other relevant or pertinent information such as languages skills, professional societies, computer literacy (list software), certifications, and appropriate outside interests.

B.   The Web Resume

The web resume typically takes the same form as your paper resume, except you have opportunities to highlight and elaborate on experiences you want future employers to know about through hyperlinks! For example, if you spent last summer working in a particle technology lab, you can link your resume to the research group's website or to a description of your project.

It is likely that a future employer (and surfer to your site) may want to print your resume. Don't rely on browsers to print it in a professional manner; allow for it to be downloaded in portable document format (.pdf), or as a Word document (.doc) or text (.txt) file.

C.   The At-A-Glance Resume

The at-a-glance resume is sometimes called the "scannable resume" because it is simple and to the point, allowing recruiters to quickly scan the resume for key words. Some large corporations began requesting this format of resumes several

167

Beyond the
Bachelor's Degree—
Things to Think
About Midway
Through

years ago in an effort to manage the high number of submissions. Like traditional resumes, at-a-glance resumes are one page with the same sections (objective, education, etc.) as traditional resumes, but instead of several sentences of description, main points are listed. Check out Appendix F for an example.

At a glance tips for the at-a-glance resume:

 Stick to font fundamentals. Pick a standard font like Times New Roman, Courier, or Palatino. Italics, bolds, and underlines should be avoided; to make headings stand out use ALL CAPS.

 Plain is perfect. Remember the value is in the text—discard lines, graphics, or bullets.

Catch phrases and key words are good.

Only submit this form of resume when specifically requested to do so. Most at-a-glance submissions are done by e-mail.

⚠ **Everything on your resume is fair game!** Be prepared to answer questions about everything on your resume (maybe in French if you list that as a spoken language!).

I managed to get a summer internship, funding for grad school, and my current job all through networking at annual meetings/conferences for my professional society. During my junior year, I began to get heavily involved with the student chapter of my professional society. I was an officer during my junior and senior years. Not only did the university send me (all expenses paid) to Anaheim, CA, Cincinnati, OH, Indianapolis, IN, and Orlando, FL, during my time as an officer, but I got the chance to meet many influential people at these conferences. They were representatives of well-known automotive and aerospace companies . . . exactly the type of place where I wanted to work one day.

Despite (or in addition to) the academic nature of these meetings, they were also very social. Many dinners and happy hours to meet with colleagues in a relaxed setting . . . perfect for getting to know someone. My faculty advisor introduced me to a Senior Engineer with Ford Motor Company. I made a point of talking with him each time I saw him at a conference: updating him on my progress in school, the accomplishments of our student chapter, and making sure to be very inquisitive about his job and recent assignments. After expressing my interest to get a summer internship with Ford (don't be afraid to be forward!), he offered me a position. And if I wanted to continue on for graduate school, Ford would pay for that too—masters or Ph.D., it was up to me. I now work fulltime at Ford in their Product Development/Vehicle Engineering Department.

Many people suggest that I was just in the right place at the right time. There may be some truth to that, but at least I knew enough to get myself to that "right" place!

(C.C., *Materials Engineering '97, University of Florida*)

The Inside Scoop on the Engineering Interview

Your technical competency (GPA) got you the interview, but now you need to get more than your foot in the door. Because the others selected for the interview also have the stamp of technical competence, you have two goals:

Make an impact and stand out from the crowd.

168

Beyond the
Bachelor's Degree—
Things to Think
About Midway
Through

Just like you try to anticipate what a professor is going to throw on a test, you should prepare yourself for an interview by predicting not just the questions but how to respond in a way that makes the interviewer want to offer you a job on the spot.

Five questions that are frequently asked of engineering students include:

1. "So what do you know about us?"

Be prepared for your interview! Even if you think you know a ton about the company with which you are interviewing, check out its website, do a magazine or journal search for current articles and issues, run over to the business library to look up statistics on the industry, and talk to professors. Keep a list of questions you would like to ask your interviewer (see question 5). It might pull time away from other work, but you'll forget all that when you get your job offer.

2. "Please draw a free body diagram of a femur bone."

You may be a good student, but how good of an engineer are you? Besides how well you know to respond to the problem, the interviewer may be evaluating you on two other qualities. The first evaluation may be how you handle having to perform in a stressful situation. The occasional interviewer has even been known to mislead the interviewee to see if he or she has enough confidence to stick to his or her knowledge. The second test is to test if you can clearly communicate your ideas and problem-solving process to another engineer. The good news is that technical questions asked in interviews tend to be fairly easy (sophomore level).

3. "So I see here that you . . ."

Some interviewers will barely look you in the eye, and just go straight down your resume and ask you questions about past honors, jobs, and skills. To stand out, have a "beacon" or two on your resume to ensure that you will be asked about it. A beacon is something that jumps out because it has an interesting name or may seem out of place (chair of the "Digital Art in a Digital Age" exhibit), but shows that you are extremely well-rounded, multitalented, and adaptive to many environments ("You spent a semester in Chile?"). An interview means open season on your resume, so be prepared to talk in depth about anything on there.

4. "You've described a positive quality, now please describe a negative quality about yourself."

This can be a tough question if you don't see it coming. Why would you tell an interviewer exactly what you don't want them to know? The interviewer is attempting to get an idea of the type of person you are. What are your values? What are you like to work with? A short, succinct answer is the way to go, but the key to this question is to turn the "negative" quality about yourself into an asset for the company: "My mind never leaves work. I am always thinking about new ways to solve problems."

5. "That's it. Do you have any questions I can answer?"

Yes! Make an attempt to interact with your interviewer on a more individual level (How long has she been working at the company? What is his or her background?). Use this opportunity to show additional knowledge about the field ("I just started reading the book, *The Prize*. Have you read it? It gives such interesting insight on the history of the oil industry."). Reiterate your enthusiasm for the company.

Not to scare you, but some sadistic interviewers *do* try to rattle you, maybe by delving into a technical project you worked on, or by asking you to spit out a short computer code on the spot that reverses a text string. Maybe they are checking performance under pressure? Be prepared! Some interviews are love fests while others are weenie roasts.

169

Beyond the
Bachelor's Degree—
Things to Think
About Midway
Through

Little Things That Can Make a Big Impact on Your Job Interview

 Come stocked. In addition to your resume, bring a pencil, blank paper, and photographs and summaries of significant engineering work you have done (a portfolio! see page 165). See the next point for what you do with your pencil and blank paper.

 Give a planned, spontaneous presentation. Draw an explanatory (well-practiced) sketch or diagram of a project you have worked on. The interviewer(s) will almost certainly ask about past projects. Make note before you go into your interview of what information you want your interviewer to pull from your responses. Your interviewer will be impressed with your ability to easily communicate technical concepts.

 Lead parts of the discussion. Think of the interview as a discussion with one of your dad's technology-loving friends, not an interviewer-interviewee conversation. Don't babble or feel you need to fill silences. Ask questions.

 Use tech-specific lingo. But use it correctly, don't overkill, and avoid acronyms the interviewer isn't familiar with! (Double negative points.) Engineers have been hired on the spot for mentioning technical buzzwords or discussing specific trends in the area of expertise of the interviewer.

Don't Forget to Write a Thank-You Note

Sending a thank-you note to every interviewer you encounter is a good idea and shows professionalism (seems ironic—you get tortured for half an hour, then have to send a thank-you note). A standard format for every interviewer is usually OK. So many notes get sent to interviewers that they probably no longer get any fun out of comparing them. The purpose of the thank-you note is: "Hey—remember me?"

What salary should you expect? According to 2002 surveys, the average starting salaries of engineering grads are the highest of all undergrad majors: $50,182 for computer engineers, $49,055 for computer scientists, $48,890 for chemical engineers, $48,613 for electrical engineers, $45,988 for industrial engineers, $45,952 for mechanical engineers, $39,598 for environmental engineers, and $37,932 for civil engineers.

170

Beyond the
Bachelor's Degree—
Things to Think
About Midway
Through

Grad School

Four years of awesome problem sets, invigorating computer programming (and debugging), and lab reports has inspired you to stay in school forever and become the perpetual student (for a little fun, call yourself a "perpetual student" around loved ones). Master's work typically takes approximately 1 to 3 years depending on the institution and whether it requires a thesis. Doctoral work is often a continuation of the research you did for your master's and can take up to an additional 3 to 7 years after your master's degree (again depending on what school you go to and what you are researching).

While many students go straight from undergrad to graduate school to do a master's and possibly a Ph.D., just as many take time off before heading back to the academic setting. If you aren't certain about what area interests you or even sure what degree you want, the best thing to do is to take time off. Get a job, climb a mountain, maybe both. Too many folks find themselves in graduate school burned out and frustrated by the lowly graduate-student financial situation.

So if you decide to be grad school bound (straight or not so straight out of undergrad school), start early!

Scads of Grad School Guidance

 Talk to a professor who can give you good advice (knows the area in which you are interested). Even after reading a dozen webpages, it can be really difficult to find the schools that do good research and work in the area in which you want to specialize. Some of the least likely schools can have the best programs for what you are interested in. Professors are very well connected in their fields through societies, conferences, and meetings. They can help you find the school that will best match your interests and give you contacts.

 Spend time researching schools. The best time to look around is the summer before your senior year. Surf potential school websites for requirements, classes, overviews of faculty and current student research activities, and so on. If you can, visit the schools and talk to current grad students. Besides the academic strengths and pressures, take note of quality of life and cost of living.

171

Beyond the
Bachelor's Degree—
Things to Think
About Midway
Through

Apply for fellowships and grants. Most engineering grad schools find funding for their grad students, but some do not. Look into and apply for independent funding offered by foundations (such as the Whitaker Foundation for biomechanics) or the government (such as the National Science Foundation or the Department of Defense). Independent funding makes you much more attractive to potential graduate schools and you will have fewer financial concerns. Your advisor or department administrator can usually direct you to a professor who is familiar with available fellowships. The Web is also an excellent source for scholarship, fellowship, and grant information. A good website to start with is FinAid, the Financial Aid Information Page, sponsored by the National Association of Student Financial Aid Administrators:

www.finaid.org/otheraid/grad.phtml

Get research experience. Most undergrads don't do research, which makes getting experience difficult (because you might have to prove yourself) and at the same time easy (because you are probably one of few undergrads interested). What does it entail? Turn to Chapter 8!

Get teaching assistant (TA) experience. If you think you may be looking for funding from your grad school, having TA experience may help you land a teaching assistantship. TAing also really helps you learn the material inside out and allows you to assist/torture others the way you were once assisted/tortured.

Study for the GRE. Get out your number 2 pencil! Or if you prefer, take the real-time computer test. Yes, we gifted left-brainers take the GRE (Graduate Record Examination) with the psychology, art history, and Spanish majors who are also going to grad school. It is the SAT revived in graduate form! There are three sections: quantitative (math), verbal, and analytical writing. The quantitative and verbal sections are worth 800 points each, and the analytical section is worth 6 points. The math section is rumored to be easier than the SAT and the verbal section is supposedly somewhat similar. The GRE recently changed from fun logic games to two writing assignments: a 45-minute "Present your perspective on an issue" task and a 30-minute "Analyze an assignment" task. You should definitely buy a GRE study guide and explore online resources—scores from these tests demonstrate how well you test, not how smart you are. If you did well on the SAT, you will do well on the GRE. If you wish you had done better on the SAT, you will do better on the GRE—just spend some time studying.

Every year, approximately 65,000 engineers graduate with their bachelor's degrees, 30,000 with their master's, and 6,000 with Ph.D.s in the United States.

Test preparation materials are included with your GRE registration. For more information and resources see: **www.gre.org** (then click on "Test preparation" for the "General test").

- *Pinpoint what you want.* Consider what this advanced degree will really give you. Do you *really* want to do more problem sets? Graduate programs are very different from school to school. Some programs can be completed in a year, do not require a thesis, and have only a specified number of classroom units in a concentration of your choice. A master's program with a thesis almost always takes longer because students are working at their own pace instead of the school's. If you think you may be interested in getting a Ph.D., be aware that students usually stay at the same school (same research project, but extended or more in-depth). So choose your school and program wisely.

- *Follow your instincts when you choose your school.* You will choose well!

The Part of the Application That Always Gets Left Until Last: The Essay!
Rumor has it that graduate schools use the essay primarily to check out your competence with the English language. There! That should take the pressure off. What gets you into graduate schools are your recommendations, grades, and GRE scores. The admissions committee (or individual) that reads your essay is attempting to get a sense of who you are by learning what important or beneficial engineering work you have done, what you are interested in studying further and/or researching, and why you want to go to that school. A good outline to follow in your essay if you don't have any better ideas is:

1. Something that tells the school about you and how you became an engineer.
2. Why you want to go to this school (based on point 1) and what area you are interested in researching or studying further.

173

Beyond the
Bachelor's Degree—
Things to Think
About Midway
Through

3. How your assets and accomplishments directly benefit the school (based on point 2). Don't be bashful. Sell yourself!
4. How attending this school fulfills all your dreams (based on 1, 2, and some literary license).

Not-Really Engineers

Once you get a B.E. or a B.S. in an engineering field, you are an engineer—no matter how far you try to run. Lately, the most common nonengineering path for graduating engineers is the consulting world where the word "analyst" crops up in the official job title. Other not-really engineers:

 Are *sales reps* for firms that need people with technical knowledge.

 Become *technical writers*.

 Go to professional schools (*doctors, lawyers, MBAs*).

 Are often the *innovative folks* (i.e., entrepreneurs) who start up their own companies.

A surprisingly large number of engineering grads do not take "trench engineering" jobs. Engineers are a commodity to the corporate world for many reasons: first, because they are problem solvers with a good tool set of analytical skills and approaches, and second, because they are used to working late nights and long hours.

The hunting process for a nonengineering job differs little from the engineering job search, so . . . get internships! Utilize the career center and connections! Polish your communication skills!

Whichever path you take. . .

Good Luck!

Appendix A
Discipline-Specific
Engineering Societies

	Organization	Website
AAEE	American Academy of Environmental Engineers	www.enviro-engrs.org
ACerS	The American Ceramic Society	www.acers.org
ACSM	American Congress on Surveying and Mapping	www.acsm.net
ACEC	American Consulting Engineers	www.acec.org
ADDA	American Design Drafting Association	www.adda.org
AHS	American Helicopter Society	www.vtol.org
AIAA	American Institute of Aeronautics and Astronautics (AIAA)	www.aiaa.org
AIChE	American Institute of Chemical Engineers	www.aiche.org
AIMBE	American Institute for Medical and Biological Engineering	www.aimbe.org
ANS	American Nuclear Society	www.ans.org
ASAE	American Society of Agricultural Engineers	www.asae.org
ASCE	American Society of Civil Engineers	www.asce.org
ASEE	American Society for Engineering Education	www.asee.org
ASEM	American Society for Engineering Management	www.asem.org
ASHRAE	American Society of Heating, Refrigerating and Air-Conditioning Engineers	www.ashrae.org
ASME	The American Society of Mechanical Engineers	www.asme.org
ASNE	American Society of Naval Engineers	www.navalengineers.org
ASNT	The American Society for Nondestructive Testing	www.asnt.org
ASQ	American Society for Quality	www.asq.org
ASSE	American Society of Safety Engineers	www.asse.org
ASSE	American Society of Sanitary Engineering	www.asse-plumbing.org
ASTM	American Society for Testing Materials	www.astm.org
ASES	American Solar Energy Society	www.ases.org
AEI	Architectural Engineering Institute	www.aeinstitute.org
ASM	ASM International	www.asm-intl.org
AACE	Association for Advancement of Cost Engineering	www.aacei.org
ACM	Association for Computing Machinery	www.acm.org
AEE	Association of Energy Engineers	www.aeecenter.org
AISE	Association of Iron and Steel Engineers	www.aise.org
AES	Audio Engineering Society	www.aes.org
BMES	Biomedical Engineering Society	www.bmes.org
CSA	Cryogenic Society of America	www.cryogenicsociety.org
IESNA	Illuminating Engineering Society of North America	www.iesna.org
IEEE	The Institute of Electronics and Electrical Engineers	www.ieee.org
IIE	Institute of Industrial Engineers	www.iienet.org
INFORMS	Institute for Operations Research and the Management Sciences	www.informs.org
IMAPS	International Microelectronics and Packaging Society	www.imaps.org
ISA	International Society for Measurement and Control	www.isa.org
SPIE	The International Society for Optical Engineering	www.spie.org
MRS	The Materials Research Society	www.mrs.org
TMS	The Minerals Metals and Materials Society	www.tms.org
NAWIC	National Association of Women in Construction	www.nawic.org
OSA	Optical Society of America	www.osa.org
SAE	The Society of Automotive Engineers	www.sae.org
SEM	Society for Experimental Mechanics	www.sem.org
SFPE	Society of Fire Protection Engineers	www.sfpe.org
SIAM	Society for Industrial and Applied Mathematics	www.siam.org
SME	Society of Manufacturing Engineers	www.sme.org
SNAME	Society of Naval Architects & Marine Engineers	www.sname.org
SPE	Society of Petroleum Engineers	www.spe.org
SPE	Society of Plastics Engineers	www.4spe.org
SES	Standards Engineering Society	www.ses-standards.org
WEF	Water Environment Federation	www.wef.org

Key: N = not available, n/a = not applicable, G = student members are part of general membership.

E-mail	Telephone	Total Members	Student Chapters
aaee@ea.net	(410) 266-3311	2,300+	N
info@acers.org	(614) 794-5890	10,000+	37
info@acsm.net	(240) 632-9716	N	N
acec@acec.org	(202) 347-7474	5,700	0
national@adda.org	(301) 460-6875	2,000	57
staff@vtol.org	(703) 684-6777	6,000	10
custserv@aiaa.org	(800) 639-2422	30,000	140+
xpress@aiche.org	(800) AICHEME	54,500	148
info@aimbe.org	(202) 496-9660	32,000	n/a
outreach@ans.org	(708) 352-6611	13,000	51
hq@asae.org	(616) 429-0300	9,000	83
(webform on website)	(800) 548-2723	120,000+	224
membership@asee.org	(202) 331-3520	10,000	6
gene.dixon@srs.gov	(573) 341-2101	1,300	6
orders@ashrae.org	(800) 527-4723	50,000	145
infocentral@asme.org	(800) THE-ASME	125,000	1,000
asnehq@navalengineers.org	(703) 836-6727	6,000	5
(webform on website)	(614) 274-6003	10,000	n/a
cs@asq.org	(800) 248-1946	136,000+	66
customerservice@asse.org	(847) 699-2929	32,000	54
info@asse-plumbing.org	(440) 835-3040	2,800	G
service@astm.org	(610) 832-9585	35,000	n/a
ases@ases.org	(303) 443-3130	6,000	3
aei@asce.org	(703) 295-6027	3,000	12
cust-srv@asminternational.org	(800) 336-5152 or (440) 338-5151	40,000	100+
info@aacei.org	(304) 296-8444	5,500	6
acmhelp@acm.org	(800) 342-6626	80,000+	430
info@AEEcenter.org	(770) 447-5083	8,000+	G
membership@aise.org	(412) 281-6323	11,000	n/a
HQ@aes.org	(212) 661-8528	N	N
info@bmes.org	(301) 459-1999	N	N
csa@huget.com	(708) 383-6220	500	G
iesna@iesna.org	(212) 248-5000 ext. 101	8,000	23
member.services@ieee.org	(800) 678-4333	320,000	950
cs@www.iienet.org	(770) 449-0461	24,000	161
informs@informs.org	(800) 446-3676	11,000	48
imaps@imaps.org	(202) 548-4001	7,000	26
info@isa.org	(919) 549-8411	50,000+	n/a
spie@spie.org	(360) 676-3290	13,000+	13
info@mrs.org	(724) 779-3003	12,400	39
tmsgeneral@tms.org	(724) 776-9000	13,000	90
nawic@onramp.net	(800) 552-3506	6,500	G
info@osa.org	(202) 223-8130	15,000	26
customerservice@sae.org	(724) 776-4970	83,000	400+
sem@sem1.com	(203) 790-6373	3,000	8
sfpehqtrs@sfpe.org	(301) 718-2910	4,600	4
siam@siam.org	(215) 382-9800	9,000	G
service@sme.org	(800) 733-4763 or (313) 271-1500	70,000	240
ccali-poutre@sname.org	(800) 798-2188	10,000+	13
service@spe.org	(972) 952-9393	12,000	109
info@4spe.org	(203) 775-0471	25,000	103
director@ses-standards.org	(305) 971-4798	N	N
dcrilley@wef.org	(800) 888-0206 or (703) 684-2452	40,000+	55+

Appendix B
Special Interest Societies

	Organization
Ethnic societies	
AABEA	American Association of Bangladeshi Engineers & Architects
AISES	American Indian Science and Engineering Society
ACE	Association of Cuban Engineers
BDPA	Black Data Processors Associates
CAHSEE	Center for the Advancement of Hispanics in Science and Engineering Education
CESASC	Chinese-American Engineers and Scientists Association of Southern California
GEM	National Consortium for Graduate Degrees for Minorities in Engineering and Science
HAES	Haitian-American Engineering Society
NACME	National Action Council for Minorities in Engineering
NSBE	National Society of Black Engineers
SACNAS	Society for the Advancement of Chicanos and Native Americans in Science
SHPE	Society of Hispanic Professional Engineers
MAES	Society of Mexican American Engineers and Scientists
VACETS	Vietnamese Association for Computing, Engineering Technology and Science
Special interest societies and organizations	
AWC	Association for Women in Computing
E Week	National Engineers Week website
JETS	Junior Engineering Technical Society
NAWIC	National Association of Women in Construction
NCEES	National Council of Examiners for Engineering and Surveying
NOGLSTP	National Organization of Gay and Lesbian Scientists and Technical Professionals
NSPE	National Society of Professional Engineers
SWE	Society of Women Engineers
WEPAN	Women in Engineering Programs and Advocates Network
WITI	Women in Technology International
WIGSAT	Women in Global Science and Technology

Key: N = not available.

Website	E-mail	Telephone
www.aabea.org	mail@aabea.org	N
www.aises.org	info@aises.org	(505) 765-1052
a-i-c.org	N	N
www.bdpa.org	(webform on website)	(800) 727-BDPA or (301) 220-2180
www.cahsee.org	N	(202) 393-0055
www.cesasc.org	chasecw@yahoo.com	(626) 572-7021
www.nd.edu/~gem	N	(574) 631-7771
www.haiti-science.com/haes/	haes@haiti-science.com	(305) 621-1189
www.nacme.org	webmaster@nacme.org	(914) 539-4010
www.nsbe.org	info@nsbe.org	(703) 549 2207
www.sacnas.org	info@sacnas.org	(831) 459-0170
www.shpe.org	shpenational@shpe.org	(323) 725-3970
www.maes-natl.org	questions@maes-natl.org	(817) 423-4332
www.vacets.org	vacets-adcom@vacets.org	(212) 387-9679
www.awc-hq.org	info@awc-hq.org	(415) 905-4663
www.eweek.org	eweek@nspe.org	(703) 684-2852
www.jets.org	info@jets.org	(703) 548-5387
www.nawic.org	nawic@nawic.org	(817) 877-5551
www.ncees.org (licensing info: www.engineeringlicense.com)	(webform on website)	(800) 250-3196 or (864) 654-6824
www.noglstp.org	office@noglstp.org	(626) 791-7689
www.nspe.org	customer.service@nspe.org	(888) 285-NSPE or (703) 684-2800
www.swe.org	hq@swe.org	(312) 596-5223
www.wepan.org	N	N
www.witi.com	membership@corp.witi.com	(800) 334-WITI or (818) 342-9746
www.wigsat.org	N	N

Appendix C
Engineering Honor Societies

Organization	Discipline
Alpha Epsilon	Agricultural Engineering
Alpha Eta Mu Beta	Biomedical
Alpha Eta Rho	Aviation
Alpha Mu	Agricultural Mechanization
Alpha Nu Sigma	Nuclear Engineering
Alpha Phi Mu	Industrial Engineering
Alpha Pi Mu	Industrial Engineering
Alpha Sigma Mu	Material Sciences
Beta Beta Beta	Biological Sciences
Beta Tau Epsilon	Manufacturing
Chi Epsilon	Civil Engineering
Delta Nu Alpha	Transportation
Epsilon Delta Sigma	Management Engineering
Epsilon Lambda Chi	Engineering Leadership
Epsilon Pi Eta	Environmental Health Studies
Epsilon Pi Tau	Research in Technology
Eta Kappa Nu	Electrical Engineering
Kappa Theta Epsilon	Co-op
Keramos	Ceramics
Gamma Epsilon	General Engineering
Gamma Sigma Delta	Agriculture
Lambda Tau	Medical Technology
Omega Chi Epsilon	Chemical Engineering
Omega Rho	Operations Research
Omicron Kappa Pi	Architecture
Order of St. Patrick	Engineering Leadership
Phi Alpha Epsilon	Architectural Engineering
Phi Lambda Upsilon	Chemical Engineering
Phi Psi	Textiles
Pi Alpha Epsilon	Building Construction
Pi Alpha Xi	Horticulture
Pi Epsilon Tau	Petroleum Engineering
Pi Mu Epsilon	Mathematics
Pi Sigma Pi	Minorities in Engineering
Pi Tau Sigma	Mechanical Engineering
Sigma Gamma Epsilon	Geological and Mining Engineering
Sigma Gamma Tau	Aerospace and Aeronautical Engineering
Sigma Lambda Chi	Building Construction
Sigma Pi Epsilon	Plastics Engineering
Sigma Pi Sigma	Physics
Sigma Xi	Engineering and Scientific Research
Tau Alpha Pi	Engineering Technology
Tau Beta Pi	National Engineering Society
Tau Sigma Delta	Architecture and Allied Arts
Upsilon Phi Epsilon	Computer Science
Xi Sigma Pi	Forestry

For more information on engineering honor societies see your advisor.

Appendix D
Some Useful Stuff

Logarithms

$$\log_b xy = \log_b x + \log_b y$$

$$\log_b \left(\frac{x}{y}\right) = \log_b x - \log_b y$$

$$\log_b x^y = y \log_b x$$

Formulas for Common Shapes

C = circumference
A = (surface) area
V = volume

Circle

$$C = 2\,\pi\,r$$

$$A = \pi\,r^2$$

Ellipse

$$C = 2\,\pi\,\sqrt{\frac{a^2 + b^2}{2}}$$

$$A = \pi\,a\,b$$

Cylinder

$$A = 2\,\pi\,r\,(h + r)$$

$$V = \pi\,r^2\,h$$

Sphere

$$A = 4\,\pi\,r^2$$

$$V = \frac{4}{3}\,\pi\,r^3$$

Greek Alphabet

Alpha	A	α	Nu	N	ν
Beta	B	β	Xi	Ξ	ξ
Gamma	Γ	γ	Omicron	O	o
Delta	Δ	δ	Pi	Π	π
Epsilon	E	ϵ	Rho	P	ρ
Zeta	Z	ζ	Sigma	Σ	υ
Eta	H	η	Tau	T	τ
Theta	Θ	θ	Upsilon	Υ	υ
Iota	I	ι	Phi	Φ	ϕ
Kappa	K	κ	Chi	X	χ
Lambda	Λ	λ	Psi	Ψ	ψ
Mu	M	μ	Omega	Ω	ω

Dimensional Prefixes

Symbol	Prefix	Multiple
T	tera units	10^{12}
G	giga units	10^{9}
M	mega units	10^{6}
k	kilo units	10^{3}
h	hecto units	10^{2}
da	deca units	10^{1}
	(no prefix) units	10^{0}
d	deci units	10^{-1}
c	centi units	10^{-2}
m	milli units	10^{-3}
μ	micro units	10^{-6}
n	nano units	10^{-9}
p	pico units	10^{-12}
f	femto units	10^{-15}
a	atto units	10^{-18}

Periodic Table of the Elements

Atomic number → Element
Symbol
Atomic weight based on $C^{12} \cdot 12.00$
( ) denotes mass number of most stable known isotope

Category labels: Light metals · Brittle metals · Ductile metals · Low melting · Nonmetallic elements · Inert gases · Rare earth elements · Lanthanide series · Actinide series · Transuranium elements

Atomic #	Symbol	Name	Atomic weight
1	H	Hydrogen	1.00797
2	He	Helium	4.003
3	Li	Lithium	6.941
4	Be	Beryllium	9.0122
5	B	Boron	10.811
6	C	Carbon	12.01
7	N	Nitrogen	14.00
8	O	Oxygen	15.999
9	F	Fluorine	18.998
10	Ne	Neon	20.18
11	Na	Sodium	22.9898
12	Mg	Magnesium	24.305
13	Al	Aluminum	26.98
14	Si	Silicon	28.086
15	P	Phosphorus	30.97
16	S	Sulfur	32.06
17	Cl	Chlorine	35.453
18	Ar	Argon	39.948
19	K	Potassium	39.098
20	Ca	Calcium	40.08
21	Sc	Scandium	44.956
22	Ti	Titanium	47.87
23	V	Vanadium	50.942
24	Cr	Chromium	51.996
25	Mn	Manganese	54.938
26	Fe	Iron	55.847
27	Co	Cobalt	58.9332
28	Ni	Nickel	58.69
29	Cu	Copper	63.546
30	Zn	Zinc	65.39
31	Ga	Gallium	69.72
32	Ge	Germanium	72.59
33	As	Arsenic	74.32
34	Se	Selenium	78.96
35	Br	Bromine	79.904
36	Kr	Krypton	83.80
37	Rb	Rubidium	85.47
38	Sr	Strontium	87.62
39	Y	Yttrium	88.906
40	Zr	Zirconium	91.22
41	Nb	Niobium	92.906
42	Mo	Molybdenum	95.94
43	Tc	Technetium	(98)
44	Ru	Ruthenium	101.07
45	Rh	Rhodium	103.905
46	Pd	Palladium	106.4
47	Ag	Silver	107.868
48	Cd	Cadmium	112.41
49	In	Indium	114.82
50	Sn	Tin	118.71
51	Sb	Antimony	121.76
52	Te	Tellurium	127.60
53	I	Iodine	126.90
54	Xe	Xenon	131.29
55	Cs	Cesium	132.905
56	Ba	Barium	137.34
57	La	Lanthanum	137.91
58	Ce	Cerium	140.12
59	Pr	Praseodymium	140.907
60	Nd	Neodymium	144.24
61	Pm	Promethium	(145)
62	Sm	Samarium	150.36
63	Eu	Europium	151.96
64	Gd	Gadolinium	157.25
65	Tb	Terbium	158.924
66	Dy	Dysprosium	162.50
67	Ho	Holmium	164.93
68	Er	Erbium	167.26
69	Tm	Thulium	168.934
70	Yb	Ytterbium	173.04
71	Lu	Lutetium	174.97
72	Hf	Hafnium	178.49
73	Ta	Tantalum	180.948
74	W	Tungsten	183.84
75	Re	Rhenium	186.2
76	Os	Osmium	190.2
77	Ir	Iridium	192.2
78	Pt	Platinum	195.08
79	Au	Gold	196.97
80	Hg	Mercury	200.59
81	Tl	Thallium	204.38
82	Pb	Lead	207.19
83	Bi	Bismuth	208.98
84	Po	Polonium	(209)
85	At	Astatine	(210)
86	Rn	Radon	(222)
87	Fr	Francium	(223)
88	Ra	Radium	(226)
89	Ac	Actinium	(227)
90	Th	Thorium	232.038
91	Pa	Protactinium	231.04
92	U	Uranium	238.03
93	Np	Neptunium	(237)
94	Pu	Plutonium	(244)
95	Am	Americium	(243)
96	Cm	Curium	(247)
97	Bk	Berkelium	(247)
98	Cf	Californium	(251)
99	Es	Einsteinium	(252)
100	Fm	Fermium	(257)
101	Md	Mendelevium	(258)
102	No	Nobelium	(259)
103	Lr	Lawrencium	(262)

Note: For more information on the elements, log onto "Periodic Table of Elements on the Internet" at: domains.twave.net/domain/yinon/default.html.

Trigonometry—Right Angled Triangle

1. $\dfrac{\text{Opposite Side}}{\text{Hypotenuse}} = \text{Sine } \theta \qquad \dfrac{O}{H} = \sin \theta$

4. $\dfrac{\text{Adjacent Side}}{\text{Opposite Side}} = \text{Cotangent } \theta \qquad \dfrac{A}{O} = \cot \theta$

2. $\dfrac{\text{Adjacent Side}}{\text{Hypotenuse}} = \text{Cosine } \theta \qquad \dfrac{A}{H} = \cos \theta$

5. $\dfrac{\text{Hypotenuse}}{\text{Adjacent Side}} = \text{Secant } \theta \qquad \dfrac{H}{A} = \sec \theta$

3. $\dfrac{\text{Opposite Side}}{\text{Adjacent Side}} = \text{Tangent } \theta \qquad \dfrac{O}{A} = \tan \theta$

6. $\dfrac{\text{Hypotenuse}}{\text{Opposite Side}} = \text{Cosecant } \theta \qquad \dfrac{H}{O} = \csc \theta$

Fundamental relations

7. $\sin \theta = \theta - \dfrac{\theta^3}{3!} + \dfrac{\theta^5}{5!} - \dfrac{\theta^7}{7!} - \dfrac{\theta^9}{9!} \cdots$

8. $\cos \theta = 1 - \dfrac{\theta^2}{2!} + \dfrac{\theta^4}{4!} - \dfrac{\theta^6}{6!} + \dfrac{\theta^8}{8!} \cdots$

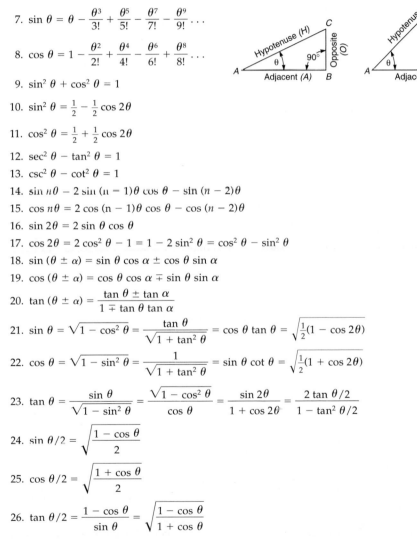

9. $\sin^2 \theta + \cos^2 \theta = 1$

10. $\sin^2 \theta = \dfrac{1}{2} - \dfrac{1}{2} \cos 2\theta$

11. $\cos^2 \theta = \dfrac{1}{2} + \dfrac{1}{2} \cos 2\theta$

12. $\sec^2 \theta - \tan^2 \theta = 1$

13. $\csc^2 \theta - \cot^2 \theta = 1$

14. $\sin n\theta - 2 \sin (n - 1)\theta \cos \theta - \sin (n - 2)\theta$

15. $\cos n\theta = 2 \cos (n - 1)\theta \cos \theta - \cos (n - 2)\theta$

16. $\sin 2\theta = 2 \sin \theta \cos \theta$

17. $\cos 2\theta = 2 \cos^2 \theta - 1 = 1 - 2 \sin^2 \theta = \cos^2 \theta - \sin^2 \theta$

18. $\sin (\theta \pm \alpha) = \sin \theta \cos \alpha \pm \cos \theta \sin \alpha$

19. $\cos (\theta \pm \alpha) = \cos \theta \cos \alpha \mp \sin \theta \sin \alpha$

20. $\tan (\theta \pm \alpha) = \dfrac{\tan \theta \pm \tan \alpha}{1 \mp \tan \theta \tan \alpha}$

21. $\sin \theta = \sqrt{1 - \cos^2 \theta} = \dfrac{\tan \theta}{\sqrt{1 + \tan^2 \theta}} = \cos \theta \tan \theta = \sqrt{\dfrac{1}{2}(1 - \cos 2\theta)}$

22. $\cos \theta = \sqrt{1 - \sin^2 \theta} = \dfrac{1}{\sqrt{1 + \tan^2 \theta}} = \sin \theta \cot \theta = \sqrt{\dfrac{1}{2}(1 + \cos 2\theta)}$

23. $\tan \theta = \dfrac{\sin \theta}{\sqrt{1 - \sin^2 \theta}} = \dfrac{\sqrt{1 - \cos^2 \theta}}{\cos \theta} = \dfrac{\sin 2\theta}{1 + \cos 2\theta} = \dfrac{2 \tan \theta/2}{1 - \tan^2 \theta/2}$

24. $\sin \theta/2 = \sqrt{\dfrac{1 - \cos \theta}{2}}$

25. $\cos \theta/2 = \sqrt{\dfrac{1 + \cos \theta}{2}}$

26. $\tan \theta/2 = \dfrac{1 - \cos \theta}{\sin \theta} = \sqrt{\dfrac{1 - \cos \theta}{1 + \cos \theta}}$

Specific Gravities and Specific Weights

Material	Average Specific Gravity	Average Specific Weight lb$_f$/ft^3	Material	Average Specific Gravity	Average Specific Weight lb$_f$/ft^3
Acid, sulfuric, 87%	1.80	112	Iron, gray cast	7.10	450
Air, S.T.P.	0.001293	0.0806	Iron, wrought	7.75	480
Alcohol, ethyl	0.790	49	Kerosene	0.80	50
Aluminum, cast	2.65	165			
Asbestos	2.5	153	Lead	11.34	710
Ash, white	0.67	42	Leather	0.94	59
Ashes, cinders	0.68	44	Limestone, solid	2.70	168
Asphaltum	1.3	81	Limestone, crushed	1.50	95
Babbitt metal, soft	10.25	625	Mahogany	0.70	44
Basalt, granite	1.50	96	Manganese	7.42	475
Brass, cast-rolled	8.50	534	Marble	2.70	166
Brick, common	1.90	119	Mercury	13.56	845
Bronze, 7.9 to 14% S$_n$	8.1	509	Monel metal, rolled	8.97	555
Cedar, white, red	0.35	22	Nickel	8.90	558
Cement, portland, bags	1.44	90	Oak, white	0.77	48
Chalk	2.25	140	Oil, lubricating	0.91	57
Clay, dry	1.00	63			
Clay, loose, wet	1.75	110	Paper	0.92	58
Coal, anthracite, solid	1.60	95	Paraffin	0.90	56
Coal, bituminous, solid	1.35	85	Petroleum, crude	0.88	55
Concrete, gravel, sand	2.3	142	Pine, white	0.43	27
Copper, cast, rolled	8.90	556	Platinum	21.5	1330
Cork	0.24	15	Redwood, California	0.42	26
Cotton, flax, hemp	1.48	93	Rubber	1.25	78
Copper ore	4.2	262			
Earth	1.75	105	Sand, loose, wet	1.90	120
			Sandstone, solid	2.30	144
Fir, Douglas	0.50	32	Seawater	1.03	64
Flour, loose	0.45	28	Silver	10.5	655
Gasoline	0.70	44	Steel, structural	7.90	490
Glass, crown	2.60	161	Sulfur	2.00	125
Glass, flint	3.30	205	Teak, African	0.99	62
Glycerine	1.25	78	Tin	7.30	456
Gold, cast-hammered	19.3	1205	Tungsten	19.22	1200
Granite, solid	2.70	172	Turpentine	0.865	54
Graphite	1.67	135			
Gravel, loose, wet	1.68	105	Water, 4° C	1.00	62.4
			Water, snow, fresh fallen	0.125	8.0
Hickory	0.77	48	Zinc	7.14	445
Ice	0.91	57			

Note: The value for the specific weight of water that is usually used in problem solutions is 62.4 lb$_f$/ft^3 or 8.34 lb$_f$/gal.

Electronics Symbols

Resistor—General	Pushbutton	Semiconductor diode
Tapped	Circuit closing Circuit opening	Undirectional diode
Variable	Selector or multi-position switch	PNP Transistor
Capacitor—General		NPN Transistor
Variable	Connector	Photovoltaic transducer
Feed-through	Jack—female contact	
Antenna—General	Plug—male contact	Fuse
	Communication connector	Lightning arrester
Dipole Loop Counter poise	2 conductor jack	Circuit breaker
	2 conductor plug	
Battery One cell Multicell	Power supply connector	Bell
Permanent magnet	Female	Buzzer
Pickup or head Writing Reading Erasing	Male	Loudspeaker
Piezoelectric crystal unit	Inductor—general	Microphone
Thermocouple (temperature measuring)	Magnetic core	Handset
	Tapped	Earphone
Conductive path or conductor	Adjustable	Headset
Air or space path	Transformer General Magnetic core	
Crossing—not connected		Fluorescent lamp (4 terminal)
Junction of paths or conductors	Electron tube components	Glow lamp (AC type)
Shielded single conductor	Directly heated (filamentary) cathode	Incandescent lamp
2 conductor cable	Indirectly heated cathode	Signal or indicator light
Coaxial cable	Cold cathode	
Ground Earth Chassis	Photo cathode	Ammeter
	Grid	Voltmeter
	Deflecting electrodes	Wattmeter
Basic contact assemblies	Anode or plate	
Close contact (break)	Applications	Generator
Open contact (make)	Triode Pentode	Motor
		Winding connections (motors & generators)
Switch Single throw Double throw	Gas filled voltage regulator Phototube	1 Phase
	Cathode ray tube	2 Phase
		3 Phase wye
		4 Phase delta

Guide to Fabric Care Symbols

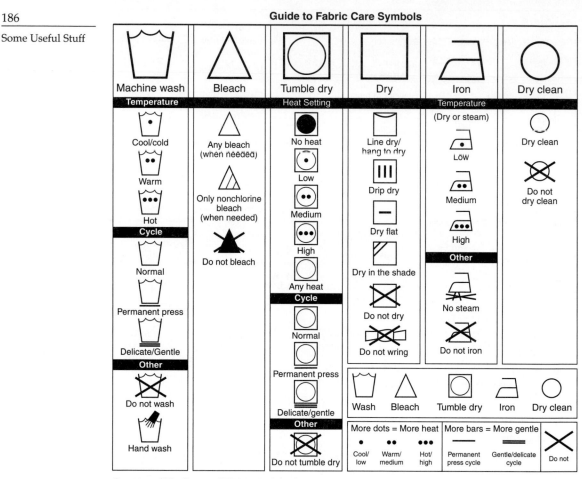

Machine wash	Bleach	Tumble dry	Dry	Iron	Dry clean
Temperature		Heat Setting		Temperature	
Cool/cold	Any bleach (when needed)	No heat	Line dry/ hang to dry	(Dry or steam) LOW	Dry clean
Warm	Only nonchlorine bleach (when needed)	Low	Drip dry	Medium	Do not dry clean
Hot		Medium	Dry flat	High	
Cycle	Do not bleach	High	Dry in the shade	**Other**	
Normal		Any heat	Do not dry	No steam	
Permanent press		**Cycle** Normal	Do not wring	Do not iron	
Delicate/Gentle		Permanent press			
Other		Delicate/gentle			
Do not wash		**Other**			
Hand wash		Do not tumble dry			

Wash	Bleach	Tumble dry	Iron	Dry clean

More dots = More heat			More bars = More gentle		
• Cool/ low	•• Warm/ medium	••• Hot/ high	— Permanent press cycle	= Gentle/delicate cycle	✕ Do not

Courtesy of The Soap and Detergent Association.

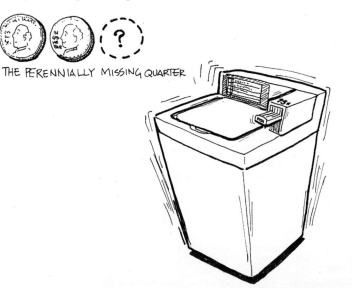

THE PERENNIALLY MISSING QUARTER

Calculus

Differential calculus formulas

1. $\dfrac{d}{dx}(x) = 1$

2. $\dfrac{d}{dx}(a) = 0$

3. $\dfrac{d}{dx}(u \pm v \mp w \pm \cdots$

$\quad - \dfrac{du}{dx} \pm \dfrac{dv}{dx} \mp \dfrac{dw}{dx} \cdots$

4. $\dfrac{d}{dx}(au) = a\dfrac{du}{dx}$

5. $\dfrac{d}{dx}(uv) = u\dfrac{dv}{dx} + v\dfrac{du}{dx}$

6. $\dfrac{d}{dx}\left(\dfrac{u}{v}\right) = \dfrac{v\dfrac{du}{dx} - u\dfrac{dv}{dx}}{v^2}$

7. $\dfrac{d(u^n)}{dx} = nu^{n-1}\dfrac{du}{dx}$

8. $\dfrac{dy}{dx} = \dfrac{1}{\dfrac{dx}{dy}} = \dfrac{\dfrac{dy}{dv}}{\dfrac{dv}{dx}}$

9. $\dfrac{d}{dx}\log_a u = \dfrac{\log_a \epsilon}{u}\dfrac{du}{dx}$

10. $\dfrac{d}{dx}\ln u = \dfrac{1}{u}\dfrac{dv}{dx}$

11. $\dfrac{d}{dx}a^u = a^u \ln a \dfrac{du}{dx}$

12. $\dfrac{d}{dx}e^u = e^u\dfrac{du}{dx}$

13. $\dfrac{d}{dx}u^v = vu^{v-1}\dfrac{du}{dx} + u^v \ln u \dfrac{dv}{dx}$

14. $\dfrac{d}{dx}\sin u = \cos u \dfrac{du}{dx}$

15. $\dfrac{d}{dx}\cos u = -\sin u \dfrac{du}{dx}$

16. $\dfrac{d}{dx}\tan u = \sec^2 u \dfrac{du}{dx}$

17. $\dfrac{d}{dx}\cot u = -\csc^2 u \dfrac{du}{dx}$

18. $\dfrac{d}{dx}\sec u = \sec u \tan u \dfrac{du}{dx}$

19. $\dfrac{d}{dx}\csc u = -\csc u \cot u \dfrac{du}{dx}$

20. $\dfrac{d}{dx}\sin^{-1} u = \dfrac{1}{\sqrt{1 - u^2}}\dfrac{du}{dx}$

21. $\dfrac{d}{dx}\cos^{-1} u = \dfrac{1}{\sqrt{1 - u^2}}\dfrac{du}{dx}$

22. $\dfrac{d}{dx}\tan^{-1} u = \dfrac{1}{1 + u^2}\dfrac{du}{dx}$

23. $\dfrac{d}{dx}\cot^{-1} u = -\dfrac{1}{1 + u^2}\dfrac{du}{dx}$

24. $\dfrac{d}{dx}\sec^{-1} u = \dfrac{1}{\sqrt[u]{u^2 - 1}}\dfrac{du}{dx}$

25. $\dfrac{d}{dx}\csc^{-1} u = -\dfrac{1}{\sqrt[u]{u^2 - 1}}\dfrac{du}{dx}$

Note: u and v = functions of x
$\quad a, e,$ and n = constants.

Appendix E
Sample Resume

Penny Gadget
Box 7065, Graceland, TN 94309
(555) 497-9699
gadget@widget.acme.edu
www.acme.edu\students\pgadget

OBJECTIVE   An intern position in water management and conservation.

EDUCATION

8/99–present   **Acme University,** Graceland, TN
B.E. in Environmental Engineering, 3.2/4.0, Expected graduation 6/03

8/95–5/99   **Seacrest High School,** Sanibel, CA
Graduated Third in class, 3.6/4.0 GPA

HONORS

1995   **Seacrest High School Student-Athlete of the Year**

ACTIVITIES

1999–present   **American Society of Civil Engineers**

2000–present   **Student Newspaper "American Acme,"** *Science Editor,*
weekly column "Ask the Science Bug"

2000–present   **Helping Hands,** Student-run community service group that assists
in local homeless shelters.

1995–1999   **Seacrest Biology Club,** President (1995–96), Sanibel, CA

1995–1999   **Seacrest Varsity Tennis Team,** Captain (1995–96), Sanibel, CA

EXPERIENCE

6/00–9/00   *Environmental Engineering Intern,* **Eastbay Water Management,** Eastbay, CA
Assisted professional engineers with review of surface water permits.
Performed site checks and investigated permit infractions.

9/99–3/00   *Research Assistant,* **Acme University Department of Earth Sciences,** Graceland, TN
Assisted graduate students with data collection and analysis.
Cataloged journal articles.

5/99–8/99   *Tennis Instructor,* **Drummond Family Tennis Centre,** Sanibel, CA
Taught lessons daily to adults and children. Organized and ran two
statewide tournaments.

COMPUTER SKILLS   C, Matlab, HTML, and Autocad
PC and Macintosh literate

REFERENCES   Available on request

Appendix F
Sample At-A-Glance Resume

Penny Gadget
Box 7065, Graceland, TN 94309
(555) 497-9699
gadget@widget.acme.edu
www.acme.edu\students\pgadget

OBJECTIVE   An intern position in water management and conservation.

EDUCATION

Acme University, Graceland, TN
B.E., Major: Environmental Engineering
GPA to date: 3.2/4.0
Expected grad date: 6/03

COURSES   Fluid Dynamics, Environmental Planning Methods, Air Pollution, Aquatic
Chemistry and Biology

SKILLS   C, Matlab, HTML, and Autocad; PC and Macintosh literate

EXPERIENCE

6/00–9/00   Eastbay Water Management   Eastbay, CA
Intern
Assisted with review of surface water permits
Performed site checks
Investigated permit infractions

9/99–3/00   Acme University Department of Earth Sciences   Graceland, TN
Research Assistant
Assisted with data collection and analysis
Cataloged journal articles

ACTIVITIES   Member of ASCE (American Society of Civil Engineers)
Science Editor for Student Newspaper "American Acme"
Volunteer at Helping Hands

REFERENCES   Available on request

Bibliography— Thanks for the Info!

Chapter 1

Digest of Education Statistics 1999 (2000), National Center for Education Statistics, U.S. Department of Education Office of Educational Research and Improvement, NCES 2000 031, p. 340.

Chapter 2

Hewlett Packard Museum online. "The Museum of HP calculators: What is RPN?": www.hpmuseum.org/rpn.htm

National Society for Professional Engineers (NSPE) website.

Chapter 3

Gough, Paddy W. "The Female Engineering Student: What Makes Her Tick?" In *Proceedings of the Conference on Women in Engineering, June 22 to 25, 1975.* Ithaca, N.Y.: Cornell University, 1975.

Holmstrom, Engin Inel. "The New Pioneers . . . Women Engineering Students." In *Proceedings of the Conference on Women in Engineering, June 22 to 25, 1975.* Ithaca, N.Y.: Cornell University, 1975.

Matyas, Marsha Lakes, and Shirley M. Malcom, eds. *Investing in Human Potential: Science and Engineering at the Crossroads.* Washington, D.C.: American Association for the Advancement of Science, 1991.

National Research Council. *Engineering Education: Designing an Adaptive System.* Washington, D.C.: National Academy Press, 1995.

O'Brannon, Helen. "The Social Scene: Isolation and Frustration." In *Proceedings of the Conference on Women in Engineering, June 22 to 25, 1975.* Ithaca, N.Y.: Cornell University, 1975.

Ott, Mary Diederich. "Attitudes and Experiences of Freshman Engineers at Cornell." In *Proceedings of the Conference on Women in Engineering, June 22 to 25, 1975.* Ithaca, N.Y.: Cornell University, 1975.

Ott, Mary Diederich, and Nancy A. Reese, eds. *Women in Engineering . . . Beyond Recruitment.* Proceedings of the Conference, Cornell University, Ithaca, June 22 to 25, 1975.

Reamon, Derek. Research presentation, Stanford University, 1998.

Shields, Charles J. *Back in School: A Guide for Adult Learners.* Hawthorne, N.J.: Career Press, 1994.

Siebert, Al, and Bernadine Gilpin. *The Adult Student's Guide to Survival and Success,* 3rd ed. Portland, Ore.: Practical Psychology Press, 1996.

Chapter 4

Aberle, Kathryn B., ABET Associate Director. "Selecting an Engineering Program in the United States," http://www.abet.org/ABET_FAQ.htm

The Iron Ring: The Ritual of the Calling of the Engineer/Les Rites D'Engagement de L'Ingenieur webpage: http://www.ironring.ca/

National Academy of Engineering, 2000. Website: "Greatest Engineering Achievements of the 20th Century," http://greatachievements.org/greatachievements/

Order of the Engineer website: http://www.order-of-the-engineer.org/

Petrosky, Henry. 1995. "The Iron Ring," *American Scientist*, Vol. 38. May/June: 229–231.
Vanderbilt University. Vanderbilt University Undergraduate Catalog Bulletin, 1994–95.

Chapter 5

Kraft, E. M., and R. E. Thomas. "Design of an Ergonomics Program for a Multifaceted
Public University," Auburn University. American Industrial Hygiene Association
1997, updated 06/27/00.
Vanderbilt University. Fall 1992 "CE 180 syllabus" (modeled after).

Chapter 6

Brown, John Fiske. *A Student Guide to Engineering Report Writing*. Solana Beach, Calif.:
United Western Press, 1985.
Faste, Rolfe. Class notes from ME 116B. (Mind map information). Stanford University,
Winter 1997.
Harvill, Lawrence R., and Thomas L. Kraft. *Technical Report Standards: How to Prepare
and Write Effective Technical Reports*. Forest Grove, Ore.: M/A Press, 1978.
Holman, J. P., and W. J. Gajda, Jr. *Experimental Methods for Engineers,* 5th ed. New York:
McGraw-Hill, 1989.
Robinson, Adam. *What Smart Students Know: Maximum Grades, Optimum Learning,
Minimum Time*. New York: Crown Publishers, 1993.

Chapter 7

Robinson, Adam. *What Smart Students Know: Maximum Grades, Optimum Learning,
Minimum Time*. New York: Crown Publishers, 1993.

Chapter 8

Alward, Barb. "Co-ops and Internships in Engineering: Learning the Ways of the
Corporate World," Diversity/Careers in Engineering and Information Technology
Summer/Fall 2000 Minority Issue, Vol. VIII, No. 3, pp. 44–52.
National Science Foundation. "Research Experience for Undergraduates—
Supplements and Sites," Program Announcement, NSF 00-107, June 13, 2000.
Penn State Co-operative Education website: http://www.engr.psu.edu/coop/ © 1999.
Resnick, Marc L., Martha A. Centeno, and Ronald Giachetti. "Research Experience for
Undergraduates—Motivating and Retaining Bright Engineering Students,"
Proceedings from the Human Factors and Ergonomics Society 2000, San Diego,
CA, July–August, 2000.

Chapter 9

Columbia University. Go Ask Alice, website, Student Health Center, Columbia
University.
Duke University. Healthy Devil website, Student Health Center, Duke University.
Huffman, Karen, Mark Vernoy, Barbara Williams, and Judith Vernoy. *Psychology in
Action,* 2nd ed. New York: John Wiley & Sons, 1991.

Chapter 10

Amell, T. K., and S. Kumar. "Cumulative Trauma Disorders and Keyboarding Work,"
1999.
Bridger, R. S., and R. S. Whistance. University of Cape Town Medical School and
Groote Schuur Hospital, "Posture and Postural Adaptation in the Workplace,"
1999.
Franklin, Benjamin. *The Autobiography of Benjamin Franklin*. Cambridge, Mass.:
Houghton Mifflin, 1928.

McTutor's History of Mathematics website, http://www-groups.dcs
.st-and.ac.uk/~history/.

Pepper, Erik, and Katherine H. Gibney, "Computer Related Symptoms: A Major
Problem for College Students," San Francisco State University. © 2000 Sandhills
Publishing Company.

Root-Berstein, Robert Scott. "Visual Thinking: The Art of Imagining Reality."
Transactions of the American Philosophical Society 75 (Part 6), 1985.

Thompson, Jennifer. "Ergonomics and College Life: Review of Research,"
http://www.louisville.edu/~jlthom02/.

Chapter 11

ETS, "Frequently Asked Questions About the General Test,"
www.gre.org/faqnew.html, 2003.

National Society of Professional Engineers (NSPE) website, http://www.nspe.org/
lc-home.htm.

Reaves, Joe. "Scannable Resumes: The High Tech Way to Get a Job," BrassRing
Campus Careers and the Technology Undergrad, Fall 2000, Vol.12,
Number 12, p. 14.

Stanford Career Planning and Placement Center. Resume handouts.

Stanford University. "Career Steps" Handout, Ambidextrous Thinking ME 313
(Engineering interviews), Fall 1995.

Appendix C

Holman, J. P., and W. J. Gajda, Jr. *Experimental Methods for Engineers,* 5th ed. New York:
McGraw-Hill, 1989.

Factoids

ASEE, *Profiles of Engineering + Engineering Technology Colleges,* 2001 edition, 2002.

Begley, Sharon. "As We Lose Engineers Who Will Take Us into the Future?" *Science
Journal* in *Wall Street Journal,* June 7, 2002, B1.

Digest of Education Statistics 1999 (2000), National Center for Education Statistics, U.S.
Department of Education Office of Educational Research and Improvement, NCES
2000-031., p. 340.

Kelsch, Shawna. "U.S. College Enrollment Among Working Parents Booming."
February 4, 2003, *News-Press,* Fort Myers, FL.

Kirscher, Cindy. "National Society of Professional Engineers Publishes 2000 Income &
Salary Survey: Engineers' Salaries Increase 1% over Consumer Price Index."
June 29, 2000, Press Release NSPE.

National Academy of Engineering. "Harris Poll Reveals Pubic Perceptions of
Engineers," September 1, 1998, NAE Press Release.

Science and Engineering Indicators (2000). Chapter 4 Higher Education in Science and
Engineering.

Tomsho, Robert. "Economic Squeeze Has More Students Working Overtime." *Wall
Street Journal,* November 5, 2002.

About the Author

Krista Donaldson graduated from Vanderbilt University in 1995 with a bachelors in Mechanical Engineering. She went on to Stanford University where she discovered the machine shop and the main reason she had wanted to become an engineer—to make stuff! This took her into the Stanford Product Design (joint masters program in Art and Mechanical Engineering) where she was able to design and produce cool things—like an ice cream cone mold and a chandelier that doubles as a swing. After gaining some teaching and research experience during her masters, she decided to continue on for a Ph.D. (in M.E.), researching how to strengthen manufacturing capacity in less industrialized economies. She can be found either in California or Kenya, or more easily at krista@donaldson.net.

About the Illustrator

Originally from Seoul, Korea, Daniel Kim is a Product Designer, currently working at Handspring, Inc., where he is designing and engineering new hand-held computers. He has also worked at IDEO Product Development, a leading product design consultancy, where he worked on designing and engineering a variety of products from eyedropper bottles and cable modems to Palm V hardcases and Handspring Visors. Most recently, Daniel was the Director of Design at DoDots, Inc. where he designed user interfaces for a web-based software application. Daniel received both his B.S. in mechanical engineering and M.S. in product design from Stanford University, where he met the author.

About the Cartoon Stripper

Born and raised in Panama, Jorge Cham obtained his B.S. from Georgia Tech and his M.S. and Ph.D. from Stanford University, all in Mechanical Engineering. His cartoons about life (or the lack thereof) in grad school can be found online at www.phdcomics.com.

Index